THE DEVIL'S THRIFT SHOP

THE DEVIL'S THRIFT SHOP

STEVEN PAJAK

The Devil's Thrift Shop
Copyright © 2025 by Steven Pajak

Published by Lakefront Terror Press

Printed in the United States of America

First Edition
Trade Paperback: 978-1-966701-14-9

CONTENTS

WELCOME TO THE SHOP

"Welcome, dear reader. I've been expecting you.

"You stand at the threshold of no ordinary establish-ment. My little thrift shop exists in the spaces between—between want and need, between past and future, between the mundane world you know and something . . . alto-gether different.

"Some find us by accident, a storefront that wasn't there yesterday. Others are drawn by something they glimpse in the window, an object that seems to call their

name. *The desperate ones always find their way here eventually.*

"Every item on these shelves carries a story within its seams, its circuits, its brushstrokes. Each has passed through hands that left more than fingerprints behind. Memories. Regrets. Desires. Power.

"I am merely the caretaker of these treasures, ensuring they find their way to those who need them most —though "need" rarely aligns with "want," as my customers discover.

"The price? Always reasonable. I've never been one to overcharge. Though I should mention that the true cost is rarely reflected on the price tag.

"As you turn these pages and meet those who've wandered into my shop, remember: every purchase is a contract. Every object, a doorway. And every story begins with someone believing they've found exactly what they were looking for.

"Do enjoy browsing. I'll be here if you need me—just follow the sound of the bell."

—The Shopkeeper

"Ah, you've noticed our holiday collection. Most people overlook these ornaments—mere trinkets to brighten winter's darkness. But this purple orb? This is something altogether different.

"The Germans call it a Hexenfänger—witch catcher. See how the black tree sits suspended within the purple glow? And that tiny white light dancing between its branches? Quite mesmerizing, isn't it?

"In Alpine villages, they speak of Frau Perchta, the

Christmas witch who visits during the twelve days of Christmas to pass judgment. For the worthy, perhaps a silver coin. For the unworthy . . . she slits their bellies open and stuffs them with straw and stones.

"These ornaments were crafted to capture her spirit—to protect the household from her terrible judgment. Though sometimes, I wonder if they contain her rather than repel her.

"I see you've noticed the small crack along the bottom. Hardly noticeable, really. I'm offering it at quite a reasonable price today. Perfect for a couple's first Christmas together, wouldn't you say?

"Shall I wrap it for you?"

———————————————————————

LaMont set the musty cardboard box on the island, wrinkling his nose at the damp odor that wafted up as he pulled open the bent and torn flaps. His fingers brushed against something soft as he reached inside, pulling out a silk-spun ornament that had been worn down to its Styrofoam core by years of use. He turned it in his hands, examining the damage. "Did you get these from a thrift store?"

"No, this weird little charity shop I spotted downtown when I was getting on the train," Ryan said, reaching over to take the ornament from LaMont's hands. He held it up to the light, his face scrunching with mild disgust. "Got the whole box for five bucks. Didn't even have to haggle with the guy."

"My bargain-hunting babe," LaMont said, reaching back into the box. His fingers found another ornament, this one a glass orb covered in red, green, and gold glitter that caught the light. At least this one was in better condition. "Someone's grandma probably died, and the kids dumped all her Christmas stuff there."

"That's sad, but Grandma had terrible taste." Ryan took a few steps toward their barren tree, stretching up to hang the ornament on a random branch without much thought. "Pathetic, right?"

"No, it's sweet." LaMont turned the gaudy Santa-face ornament in his hands, his thumb tracing over its worn features. "Nothing says Christmas like other people's discarded memories. Plus, you saved like fifty bucks on decorations."

Ryan pushed away from the granite island where he'd been leaning, closing the distance between them. His lips found LaMont's in a quick kiss. "I love that you always put a positive spin on everything."

"It's what I do—bring light into our dark, second-hand-decorated world."

LaMont moved to hang his ornament next, taking

more care with its placement than Ryan had. He felt something press against his leg and looked down to find Chairman Meow winding between his feet. With a swift motion, he scooped up the cat and unceremoniously deposited him onto the sofa.

"Chairman Meow, you will stay away from this tree, you devious little bastard."

"Why do you despise my kitty, Lammie?"

LaMont bristled at the nickname but swallowed his annoyance—he loved his fiancé too much to make a fuss. He returned to the island, methodically pulling ornaments from the box and arranging them on the cool granite surface. "Because he's devious. And he's spiteful."

"He's a cat. It's in his DNA."

"If he knocks that tree down, my foot's gonna be in his ass."

Ryan laughed, his hands busy wrapping a frayed strand of tinsel around the top half of their seven-foot tree. The metallic strands caught the light as he worked, revealing worn patches and missing sections. He paused, frowning at his handiwork. "We should probably start with lights, right? I feel like we should start with lights."

"Uh-huh," LaMont said, his attention clearly elsewhere as his fingers brushed against something unexpected at the bottom of the box. "Always start with lights."

Ryan's hands stilled on the tinsel. "Why did you let me get this far?"

LaMont barely heard him. He spread the remaining ornaments across the island's surface before setting the empty cardboard box on the floor. His eyes were fixed on what he'd found. "Babe, what is this?"

Before him sat an ornate wooden box, stained a deep shade of purple that seemed to absorb the kitchen light. His fingers traced over the lid's engravings: a triple-moon symbol, carefully filled in with what appeared to be silver. The full moon dominated the center, flanked by two crescent moons. Within the center moon, a star—looking more like an upside-down pentagram—caught the light, while ivy vines spiraled outward from beneath it, reaching toward the box's edges.

Ryan crossed the room slowly, his footsteps hesitant as his eyes examined the box. "I don't know. Where did you find it?"

"In the stinky box," LaMont said, unable to tear his gaze away from the mysterious container. Finally, he looked up at Ryan.

"Well, open it," Ryan suggested.

LaMont's fingers hesitated over the hinge, then carefully worked it open. As he slowly lifted the lid, his free hand flew to his mouth. "Oh my, it's . . . beautiful."

Ryan hurried to LaMont's side, draping an arm over

his shoulder as they both leaned in to look. "Wow. It's more than beautiful. What is it?"

"I don't know."

LaMont reached for the object nestled inside. The glass orb was four inches in diameter, its top half radiating a luxurious purple hue. It sat perfectly in a precisely formed depression, protected by deep purple felt lining. When his fingers made contact, he startled at its coolness—he'd expected warmth, though he couldn't say why. A glass loop, about a half inch around, protruded from the top, its purpose unclear. The orb made a soft sucking sound as he lifted it from its protective bed, and his eyes widened as he cradled the glorious piece in his palm.

Both men leaned closer, their breath fogging the glass as they examined their exquisite find. Within the center of the orb stood a tree shaped of black glass, its five thick branches extending from the trunk before disappearing into the magnificent purple aura. They exchanged puzzled glances, neither understanding how the globe could radiate such light without any visible power source. As they watched, a much smaller, more brilliant orb of light, no larger than a pea, floated out from the purple radiance. It hovered near one of the branches before slowly weaving its way among the others.

"Oh my God, do you see that?" LaMont practically squealed.

"I'm seeing it."

They watched, mesmerized, as the white firefly made a circuit, changing up its pattern of flight to an infinity loop before finally disappearing into the purple haze. LaMont shook the globe gently, hoping the tiny white orb would reappear, but it did not.

"What is it?" Ryan asked again.

Setting the globe ever so gently into the box, LaMont examined the interior further. On the underside of the lid, he found a silver plaque, two inches long and one inch high, with one word etched in black: *Hexenfänger*.

LaMont turned his head slightly and looked up at Ryan. "Any idea what that means?"

Ryan shook his head.

"Hecks-en-faner? Fan-ger? Sounds German. You think it's German?"

"I have no idea."

"I'm gonna look it up."

"You do that," Ryan said. "I'm going to feed the Chairman."

LaMont pulled up the Google Translate app on his phone, carefully typing out each letter as Ryan's footsteps moved toward the kitchen sink. "It says Hexenfänger translates to 'witch catcher.'"

Ryan knelt at the sink cabinet, the hinges creaking as

he pulled out a bag of kibble. "Witch catcher? Witch as in, like . . . a hag?"

"I guess. Hold on, I'm still looking." LaMont's fingers moved across his phone screen, eyes scanning rapidly.

The sound of Chairman Meow's metal food bowl scraping across the kitchen floor filled the silence as Ryan poured out the kibble, the cat weaving impatiently between his feet.

"Okay," LaMont said, leaning forward on his elbows as he read from his phone. "It says it is an ornament designed by German master craftsmen for the purpose of capturing the spirit of Frau Perchta. In the folklore of Bavaria and Austria, the ghost of Frau Perchta, the Christmas Witch, is said to roam the countryside at midwinter and enter homes during the twelve days between Christmas and Epiphany—especially on the Twelfth Night." His voice grew more animated as he continued reading. "She would know whether the children and young servants of the household had behaved well and worked hard all year. If they had, they might find a small silver coin the next day, in a shoe or pail. If not"—his eyes widened at the next part—"she'd slit their bellies open and stuff the hole with straw and pebbles."

Ryan straightened up from feeding Chairman Meow,

his face twisting in disgust. "Jesus, that's fucking gruesome."

"Yes, it is." LaMont's gaze drifted back to the *Hexenfänger*, his fingers hovering over its surface without touching. An item of such great beauty designed to contain such ugliness. It was poetic. "The ornament is said to keep her ghostly spirit away—or, if used properly, could capture her within. Do you think this is the only one?"

"What, the ornament?"

"Yes."

Ryan scoffed, closing the cabinet door with his foot. "No. There are probably tens of thousands of those things around. Does it say 'Made in China' on the bottom?"

"Don't be silly. This is . . . exquisite." LaMont's fingers traced the ornate details. "This was forged with love by a very skilled craftsman."

"Well, it is pretty, I'll give you that." Ryan moved behind LaMont, wrapping an arm around him and resting his chin on his shoulder. "You gonna help me get these things on the tree?"

"In a minute. I want to look into this a bit more."

"You're such a nerd for that folklore shit." Ryan pressed a quick kiss to LaMont's temple. "Do your thing, boo. I'll toil away at the tree while you get your geek on."

Ryan dropped down in front of the tree, crossing his

legs as he grabbed the string of lights. After a few minutes of wrestling with the tangled mess, his frustrated growl broke the quiet. He hurled the lights across the floor, making Chairman Meow leap back with a startled yowl. "I hate those fucking things. We're not doing lights on the tree."

"Babe, come here. You gotta see this." LaMont's voice held an urgency that made Ryan look up.

"Yeah, yeah, coming." Ryan pushed himself up from the floor, his socked feet sliding slightly on the hardwood as he made his way to the island. He settled onto the stool beside LaMont. "What am I looking at?"

LaMont held his phone out, screen tilted so Ryan could see. "This looks just like ours, right?"

Ryan leaned in, squinting at the eBay listing. The ornament on the screen was nearly identical to the one they'd found in the charity shop box—except instead of the purple aura, this one radiated a crisp, stunning blue.

"Almost exactly. Is that . . ." Ryan blinked, leaning closer to the screen. "Is that current bid really $9,000?"

"Yes!" LaMont's fingers flew across the screen, scrolling through more listings. "I looked at past auctions for these, and I saw one that sold for nearly $20,000."

"Get the fuck out." Ryan's voice was barely above a whisper. "That thing was in a five-dollar discount box . . ."

Ryan scrolled through the listing, scanning the same folklore about the ornament that LaMont had read earlier. He shook his head slowly, still struggling to believe that people would pay that much, no matter how beautiful the piece was.

"Should we list it?"

LaMont turned to stare at him, his hand tightening on Ryan's. "Are you serious?"

"Yes."

"No."

"Why not?"

"Because."

"Because why?"

"Because we should keep it. Make it our family heirloom."

"Our what now?" Ryan's eyebrows shot up.

"You know, something special to pass along. To our kids someday." LaMont's fingers traced the edge of the wooden box, not meeting Ryan's eyes.

"Special? Uh, first off, it's just some random ornament we got at a charity shop for five bucks. And second"—Ryan turned on his stool to face LaMont directly—"we have no one to pass it along to. Unless you have a uterus . . ."

"We could adopt."

Ryan raised an eyebrow, studying LaMont's face. "I

thought you didn't want kids. You said kids were worse than Chairman Meow and more expensive. Are you changing your mind?"

LaMont's shoulders slumped as he sighed. "No."

"All right then." Ryan's hand found LaMont's knee. "Let's sell it. We can use the money to get new furniture or something. Or for a trip to Rome. You do want to see Rome—I know you do."

"I do, I do." LaMont paused, his fingers still absently stroking the box. "Can we at least enjoy it for the next few days? And then, after Christmas, if you still want to list it, we'll do it?"

Ryan considered it, watching LaMont's face. This was their first Christmas together—not just as a couple, but also in their first apartment. The ornament obviously made Lammie happy. He could wait a few days before selling and raking in ten grand (or more), which would make him happy.

"Go on and find a spot for it on the tree," Ryan said, standing up from his stool. "The rest of this can wait until tomorrow. Let's go fool around to celebrate our good fortune."

A LOUD THUD, followed by the sound of something rolling across the hardwood floor, jerked LaMont from sleep. The sour acid of dread pooled in his belly as he kicked off the sheets, his feet hitting the cold floor. He knew exactly what that sound was, and God help Chairman Meow if he was right.

He fast-walked down the dark hallway in his boxer shorts, his dark skin blending into the shadows. Chairman Meow streaked past him as he entered the kitchen, nearly tripping him. A moment later, when LaMont saw the *Hexenfänger* near the sink, he wished he'd stomped the fucking cat's guts out.

"No, no, no!" His knees hit the hard floor as he scooped the ornament into his hands. His fingers moved frantically over the surface, examining every inch. It appeared unscathed at first, but then his thumb found it—a crack along the bottom, about a half-inch long, with six or seven fingerlike fissures spreading from it like a spider's web.

The kitchen light flicked on, flooding the room with sudden brightness.

"What happened?" Ryan's voice was thick with sleep.

LaMont stood, his legs stiff from the cold floor. He glared at Ryan, his palms gently cupping the globe while his fingers gripped it tight. "Your fucking cat knocked it to the floor, and it's cracked!"

Ryan hurried over, his bare feet padding across the floor as he leaned in to examine the damage. "Yeah, it's cracked all right. Good thing it's on the bottom. It's not even that noticeable."

LaMont fixed Ryan with a withering stare, his voice dripping with sarcasm. "You're not concerned what it will do to the resale value?"

"Wow, you had that one locked and loaded." Ryan took a step back, running a hand through his sleep-mussed hair. "Why you pissed at me? I didn't do anything."

"You're right. It's your fucking cat. I knew he was going to do something stupid like this."

"Lammie, he's a cat. That's what they—"

"Stop calling me that!" LaMont's voice rose sharply. "You should do a better job of keeping that cat in line instead of making excuses. He fucks with this again, I'll put his ass in a shelter so fast your head will spin."

Ryan stood frozen, mouth slightly open, stunned speechless. All he could do was watch as LaMont carefully hung the ornament back on the tree and stalked past him, disappearing down the dark hallway toward their bedroom.

RYAN WOKE with a dull ache in his bladder. He fumbled for his phone on the nightstand: 4:18. His hand reached across the cool sheets, finding LaMont's side of the bed empty. After Chairman Meow's attack on the ornament, LaMont had lain there, quietly stewing until they both fell asleep. He'd been pissed, sure, but not enough to sleep in another room.

Blindly, Ryan pressed his feet into his slippers, letting out a quiet grunt as he pushed himself up from the bed. The hardwood was cold even through the thin soles as he shuffled down the hall, pausing in the bathroom to relieve himself.

At the kitchen threshold, his hand hovered over the light switch as his eyes adjusted to the darkness. LaMont stood in front of the Christmas tree, his body unnaturally still, eyes locked on the ornament.

Ryan watched him for a moment, waiting to see if LaMont would notice he was no longer alone.

He cleared his throat, the sound loud in the quiet apartment. "You standing guard?"

No response. LaMont didn't even blink.

"LaMont?"

Still nothing. Ryan crossed the kitchen, his slippers whispering against the floor. He stood beside his fiancé and put a hand on his shoulder, feeling the tension there. "Lammie?"

LaMont continued to gaze at the *Hexenfänger*, but his eyes darted as though tracking something.

Ryan followed his gaze to the ornament and saw it—the white firefly was back, swirling and looping through the purple haze. It moved more erratically than before, its flight jittery, desperate.

"Did you hear it, too?"

Ryan startled at LaMont's voice, low and dreamy, like someone speaking in a trance.

"What?"

"The sound of the firefly. It woke me."

Ryan listened, straining his ears. The familiar hum of the refrigerator. The soft whoosh of the furnace blower. Nothing else.

He took LaMont's arm, startled by how cold his skin was. "Let's go back to bed."

"It wants to get out."

Ryan frowned. "The firefly?"

LaMont nodded slowly, then suddenly his arm shot up, finger pointing at the globe. "There. It found a way."

The firefly pressed itself against the bottom of the ornament, its glow intensifying until it burned white-hot.

Threads of light seeped through the crack, impossibly escaping through the fissures in the glass.

It shouldn't have been able to.

The crack was shallow, barely reaching the inner portions of the globe. And yet, the firefly slipped through, its tiny orb reforming outside, hovering in front of them, pulsing—catching its breath.

Ryan and LaMont stood frozen as it floated closer, rising to nose level with LaMont, eye level with Ryan.

Then, before either could react, it shot left with shocking speed—so fast Ryan lost track of it—only to see it a split second later as it disappeared into LaMont's left nostril.

"What the fuck!"

LaMont stumbled backward, slamming into the kitchen island. His hands clawed at his face, fingers pinching his nose as he snorted violently, desperate to expel the firefly.

Ryan stood paralyzed, his hands frozen at the sides of his face, mouth agape as his fiancé convulsed in a frantic, jerking panic.

Then, just as suddenly, LaMont went still.

His hands dropped limply to his sides. His chest rose and fell with a deep, shuddering breath. His throat clicked audibly when he swallowed.

His eyelids fluttered. Too long. Ten seconds at least.

Then, finally, his eyes shifted to Ryan. "Babe? What's happening?"

Ryan shook his head, finally able to move. He grabbed LaMont, crushing him in a fierce hug before pushing him back to arm's length, his fingers digging into LaMont's triceps. "Are you okay?"

LaMont blinked, his face a mask of confusion as he looked around the kitchen. "What are we doing in here?"

Ryan stared at him. Was he serious?

"That . . . that thing went inside you."

LaMont glanced at the microwave clock. "It's too early to be up. Let's go back to bed."

Ryan opened his mouth to argue, but LaMont was already pulling away. His bare feet padded softly across the kitchen floor, disappearing into the dark hallway toward their bedroom.

Ryan remained standing there, his mouth still open, wondering what the hell had just happened.

THE FOLLOWING MORNING, Chairman Meow wouldn't emerge from the cat hotel that stood in the corner behind the sofa. Ryan knelt on the floor, shaking a bag of treats;

the familiar rattle was usually enough to bring his cat running. When Chairman Meow finally crept out, his movements were hesitant, tail low. The moment LaMont entered the room, the cat's entire demeanor changed—hackles raised, back arched into a sharp curve, and a low hiss escaped as he retreated back to his hiding place.

"I guess he's still upset over last night," Ryan said, the treat bag crinkling in his hand.

"That's a guilty conscience," LaMont said, the coffee maker gurgling as he poured the dark liquid into his go-cup. He grabbed his briefcase from one of the island stools, the leather creaking as he lifted it. "Make sure he keeps away from the tree while I'm gone."

"Babe, let it go."

"Nope. Chairman Meow is now my sworn enemy." LaMont's voice held an edge Ryan had never heard before.

He pressed a quick kiss to Ryan's cheek, his lips surprisingly cold, and left for work. The door clicked shut behind him with unusual force.

That evening, after dinner, they settled onto the sofa to watch television; the familiar warmth of their usual routine almost made the previous night's events feel like a dream. They were three episodes into the fourth season of The Sinner when the television and lights began to flicker, the screen stuttering with static.

"The cat is messing with the outlets back there again," LaMont growled, starting to rise, but Ryan's hand on his leg stopped him.

"No, I've got this." Ryan moved around the back of the television cabinet, his fingers finding the tangle of cables. "Any better?"

The picture cleared, and LaMont was about to respond when the face of a horrifying hag flashed across the screen. "Motherfucker!" His scream pierced the quiet apartment as he jumped, one hand clutching his chest where his heart hammered against his ribs.

"What? What happened?" Ryan's voice came muffled from behind the television.

"Did you do that?"

"Do what?" Ryan peered over the top of the television, his face creased with confusion.

LaMont stared at the screen, his eyes wide and searching, but the hag's face didn't reappear. The show played on as if nothing had happened.

"Lammie, what happened?"

"You're not playing one of those pranks on me?" LaMont's voice trembled slightly, his eyes still fixed on the screen.

Ryan emerged from behind the television, concern etched on his face. "I am so lost. What prank?"

LaMont shook his head, fingers gripping the edge of

the sofa cushion. "One of those jump-scare videos when a scary-ass face comes on screen out of nowhere. Like the one we watched on YouTube, where the kids are watching some cartoons on the computer and then, suddenly, a ghoul face pops up and they shit their pants?"

Ryan shook his head slowly. "How would I do that? Hack into Netflix?"

"Fuck, I don't know."

"What did it look like? Maybe it was part of the show?"

"No, it wasn't." LaMont's jaw clenched. "Just forget it. Was the cat fucking with the wires?"

"No, he's in his cat hotel. I think he's afraid of you now."

"Good. He should be." LaMont shot a dark glare toward the cat's hiding spot. On screen, The Sinner played on in perfect 4K clarity. He settled back against the cushions, trying to relax, and within a few minutes, he'd almost convinced himself he'd imagined it.

BEFORE BED, LaMont stood under the shower spray, turning the temperature up until steam filled the bathroom despite the vent fan whirring on high. Hot water cascaded over his shoulders, but it couldn't seem to chase away the chill that had settled in his bones. When he finally stepped out, he wrapped a towel around his waist, the fabric catching on his damp skin. His hand wiped a clear patch in the foggy mirror, and he opened the cabinet for his toothbrush. As he closed the door, he caught movement in the mirror and jumped—behind him stood the fucking hag. He whirled around, toothbrush raised like a weapon, his heart thundering in his chest.

The bathroom was empty. Steam curled through the air where the hag had been. His pulse throbbed in his ears with each rapid beat. Slowly, he stepped forward, his feet leaving wet prints on the tile as he reached for the shower curtain. With one quick motion, he yanked it all the way open. Nothing but an empty tub, water still draining with a quiet gurgle.

He let out a shuddering breath he hadn't realized he'd been holding.

Back at the sink, he squeezed a pea-sized amount of toothpaste onto his brush and started his meticulous routine, counting each stroke—up and down, side to side, front, tops, and sides. The ritual was comforting, familiar. Ever since childhood, he'd been anal about his brushing regimen, proud to boast he'd never had a single cavity.

He cupped water in his hands, swished and spat,

repeated the motion, then filled his mouth with the sharp bite of mouthwash. As he pulled back his lips to check his teeth, the mirror fogged over again. His hand moved to wipe it clear, but he yelped and stumbled backward when the hag's opaque eyes stared back at him through the mist. His calves hit the tub edge, and he fell, desperately grabbing at the sidewall and curtain before his ass landed hard on the porcelain rim.

"Motherfucker!"

He scrambled up immediately, hands frantically wiping away all moisture from the mirror's surface. His chest heaved as he stared at his own reflection, watching beads of sweat—or maybe shower water—roll down his face. The hag didn't return. Had she ever been there? What the fuck was it with the hag? He had no reason for her to be stuck in his head . . . for his mind to play these tricks.

In bed, Ryan was already sleeping soundly as LaMont slipped under the covers, exhaustion quickly pulling him under. A sound woke him hours later. His eyes opened to find the hag floating above him, near the ceiling. Her black tattered robe swayed in a nonexistent breeze, dirty frayed hair hanging like dead vines. Her face—half goat, half woman—twisted into a grotesque smile. Dead eyes bored into his soul. He tried to move, to get up, but terror paralyzed him. He tried to cry out for Ryan, but his voice was trapped in his throat. The hag drifted closer, closer still, until she was nose to nose with him, those opaque eyes filling his vision. They locked gazes, and then her lips peeled back, revealing rows of spiked teeth, and she screamed—her hot breath burning his flesh and . . .

LaMont bolted upright, his body drenched in sweat, heart racing. Chairman Meow sat perched on the corner chair, amber eyes fixed on him. A strange mewling sound emerged from deep in the cat's belly.

"Go fuck yourself, cat." LaMont hurled a pillow across the room. The cat darted away into the darkness. He barely slept the rest of the night.

THEY SPENT the morning lounging in their pajamas, mugs of hot chocolate warming their hands. Ryan had added too many marshmallows to his cup, as usual, creating a sticky mountain that made LaMont laugh.

"You're going to get diabetes," LaMont said, reaching over to swipe a marshmallow.

"Worth it." Ryan caught his hand, kissed his fingertips. "Besides, you'll take care of me if I do."

"Always."

One of the benefits of not having family around for the holidays was the freedom to sleep in or just fucking relax. Around noon, the smell of butter and maple syrup filled the apartment as they made pancakes. Ryan managed to burn his edges while LaMont's came out perfectly golden.

"How do you do that?" Ryan poked at his charred pancake with a fork. "It's like you have pancake superpowers."

LaMont hip-checked him away from the stove. "It's called patience, babe. Here, have one of mine." He slid a perfect golden disc onto Ryan's plate. "Can't have you starving on Christmas Eve."

The afternoon found them in the kitchen, Christmas music drifting from Alexa's tinny speaker while they prepped their holiday feast for two. Ryan proved himself hopeless in the kitchen, nearly slicing his finger while chopping vegetables.

"Jesus!" LaMont grabbed his hand, examining the near-miss. "Maybe you should stick to stirring things."

"But I want to help," Ryan pouted, looking so pitiful that LaMont had to kiss him.

"Fine. But if you lose a finger, I'm not taking you to the ER until after dinner."

LaMont moved with practiced ease, orchestrating their meal like a conductor. The combined aromas of roasting turkey, glazed ham, and bubbling lasagna transformed their small apartment into something that felt like home.

They settled in front of back-to-back Hallmark specials, Ryan sprawled across the couch with his head in LaMont's lap. Every few minutes, LaMont would catch himself getting misty-eyed at the predictable romance scenes.

"Are you crying?" Ryan twisted to look up at him, grinning.

"No," LaMont wiped his eyes quickly. "Just allergies."

"Uh-huh. You big softie." Ryan reached up to touch his cheek. "Don't worry, I'll never tell anyone you're a Hallmark gay."

"I will smother you with this pillow."

Their eclectic Christmas Eve dinner came together around eight, steam rising from their loaded plates. Ryan had insisted on using the "fancy" plates—a mismatched set from the charity shop that he'd fallen in love with.

"This is incredible," Ryan said around a mouthful of turkey. "How did I get so lucky?"

LaMont reached across the table, squeezed his hand. "You bought me at a charity shop, remember? Best five bucks you ever spent."

They shared comfortable silence broken only by

appreciative murmurs and the occasional footsie under the table. At nine, they each savored a slice of pecan pie with coffee, the rich sweetness a perfect end to their meal.

"This is better than my mom's," Ryan said, then quickly looked over his shoulder. "Don't tell her I said that."

"Your secret's safe with me." LaMont collected their plates, pausing to kiss the top of Ryan's head. "Though I might use it as blackmail someday."

By ten p.m., they gathered around their secondhand tree. Ryan insisted they each open one present, a tradition from his childhood. LaMont carefully unwrapped a small box to find a vintage watch he'd admired months ago in an antique store window.

"You remembered," he breathed, turning it over in his hands.

"Of course I did." Ryan fastened it around LaMont's wrist. "Now you'll always know what time it is in my heart."

"That was incredibly cheesy."

"You love it."

"I love you."

They retired to the bedroom to make Christmas love, their bodies moving together in the soft glow of string lights from the window. Ryan's fingers traced patterns on LaMont's skin, his touch full of tenderness and promise.

LaMont held him close, breathing in the familiar scent of his shampoo, wondering how he got so lucky.

Long after midnight, LaMont's eyes snapped open. Something cold moved through him, like ice water replacing his blood. His body rose from the bed without his permission, muscles responding to commands he wasn't giving. Inside his head, a voice that wasn't his own whispered in ancient German, words he shouldn't understand but did: *"Zeit für das Urteil."* Time for judgment.

He tried to fight it, to lie back down, to wrap his arms around Ryan's sleeping form. But his body wouldn't obey. His feet touched the cold hardwood, and he stood, watching himself move as if from far away. Ryan's soft breathing filled the room—in, out, peaceful and unaware.

Stop, LaMont thought. Please stop. But his legs carried him from the bedroom, each step silent and precise. The hallway stretched before him, moonlight casting strange shadows through the windows. His body moved with terrible purpose toward the kitchen, though his mind screamed in protest.

The *Hexenfänger* hung on the tree, its purple aura pulsing like a heartbeat in the darkness. LaMont stood before it, unable to look away, unable to blink. The tiny crack along the bottom seemed to whisper, releasing tendrils of ancient magic that wrapped around him like chains. Five minutes passed as he stared, unblinking, tears rolling down his cheeks while that cold voice filled his head with judgment.

"*Untauglich*," it whispered. Unworthy.

His hand moved to the knife block, fingers wrapping around the handle of the filet knife. The blade made a soft whisper as it slid free. Stainless steel gleamed dully in the moonlight, and LaMont watched in horror as his thumb tested its edge.

No, he thought. *Please, God, no.*

His feet carried him back to the bedroom, each step measured and silent. Ryan had rolled onto his back, one arm flung above his head, vulnerable and exposed. The moonlight painted stripes across his bare chest through the blinds. LaMont stood beside the bed, the knife hanging loosely at his side, and fought with everything he had against what was coming.

Twenty minutes passed as he stood there, tears streaming down his face, his body rigid with the battle raging inside him. Ryan's chest rose and fell, rose and fell, each breath precious and numbered. The voice in LaMont's head grew louder, demanding justice, demanding punishment. His arm began to rise, the knife catching moonlight.

Please, he begged silently. Not him. Take me instead.

The voice laughed, cold and ancient. "*Das ist deine Strafe.*" This is your punishment.

His arm rose higher, and in that moment, Ryan's eyes fluttered open. He blinked sleepily, a soft smile forming as he saw LaMont standing there. The smile died as his eyes adjusted to the darkness, taking in LaMont's rigid posture, the tears on his face, the knife held high.

"Lammie?" Ryan's voice was small, confused. "What are you—"

The knife plunged down with devastating force. Ryan's scream pierced the night as the blade sank deep into his abdomen. His hands flew up, grabbing LaMont's wrists, trying to stop what was already done. Blood, hot and slick, poured over their joined hands.

"Please," Ryan gasped, his eyes locked on LaMont's face. "Baby, please stop—"

But LaMont's face wasn't his own anymore. His eyes had gone milky white, and when he spoke, an ancient voice emerged: "*Untauglich. Unrein. Unwürdig.*" Unworthy. Unclean. Unfit.

LaMont felt his body climb onto the bed, straddling Ryan's knees. His hands gripped the knife with terrible purpose, and in one forceful yet fluid motion—guided by Frau Perchta's centuries of practice—he tore the blade backward, splitting Ryan open from navel to groin.

Ryan's scream became a wet gurgle. His hands scrambled at his stomach, desperately trying to hold himself together as LaMont rose from the bed. The knife clattered to the floor. Blood soaked into the sheets, spreading in a dark pool beneath Ryan's writhing body.

"Help," Ryan gasped, grabbing a pillow and pressing it against the gaping wound. "Oh God, Lammie, please help me . . ."

But LaMont's body moved with mechanical purpose toward the door. Behind him, he could hear Ryan's labored breathing, punctuated by wet, painful sobs. In the kitchen, his hands began opening cabinets and drawers. They found uncooked rice from the pantry, dried herbs from their spice rack, the decorative moss from their fall centerpiece. His fingers crushed dried bay leaves and thyme, mixing them with the rice in a grotesque parody of stuffing.

Next, the rock garden. The small decorative bowl on their coffee table held smooth river stones they'd collected on their first vacation together. LaMont's hands scooped them up, the rocks clicking together musically.

When he returned to the bedroom, Ryan had managed to drag himself halfway off the bed, leaving a wide smear of blood across the sheets. The pillow, now soaked crimson, was still clutched against his stomach.

"Baby," Ryan whimpered, seeing LaMont in the doorway. "Please call . . . please . . ."

But LaMont's body moved forward with terrible purpose, kneeling beside Ryan. His hands, gentle as if preparing a holiday meal, began their work. Ryan screamed as LaMont pulled the pillow away and started packing the wound with rice and crushed herbs, pushing the mixture deep inside. The river stones followed, each one placed with ritualistic precision.

"Stop," Ryan begged, his voice growing weaker. "It hurts . . . Lammie, why . . ."

"*Sei still*," the ancient voice emerged from LaMont's mouth as his hands continued their terrible work. "Be still."

Ryan's fingers clawed weakly at LaMont's arms, leaving bloody trails across his skin. Each stone disappeared into the cavity—stones they'd collected from that beach in Oregon, the ones Ryan had insisted on keeping because they were "perfect skipping stones." Now they vanished one by one into his body, alongside grains of rice and fragments of herbs that stuck to the blood coating LaMont's fingers.

Inside his mind, LaMont screamed as he recognized each stone: the heart-shaped one from their first morning on the beach, the speckled gray one that Ryan said looked like Chairman Meow, the flat black one they'd used as a coaster all summer. His hands kept moving, mechanical and precise, while tears streamed down his face.

Ryan's struggles grew weaker. His eyes, glazed with pain and confusion, locked onto LaMont's face. "I don't . . . understand . . ." he gasped between shallow breaths. "I love . . ."

Blood bubbled at the corners of his mouth. His hand reached up one last time, touching LaMont's cheek with trembling fingers. The touch was gentle, loving even now, and that made it so much worse. LaMont felt the exact moment Ryan's hand went slack, watched as the light in those familiar eyes dimmed and faded.

Then, like a rubber band snapping, the cold presence released him. The glowing orb erupted from his mouth in a violent burst, bringing with it the taste of ancient winters and bitter judgment. It hovered before him, pulsing with satisfied light, illuminating the horror he'd created. In its glow, he could see every detail with perfect clarity: the rice and herbs mixing with blood and other things he didn't want to name, the stones nestled obscenely in Ryan's flesh, the absolute stillness of the body before him.

The orb drifted away, moving with deliberate slowness toward the Christmas tree, as if wanting LaMont to fully appreciate its handiwork. It disappeared into the crack in the *Hexenfänger* with a soft whisper, like a satisfied sigh. The purple aura flared briefly, then settled back to its usual gentle glow.

In the sudden darkness, LaMont became aware of three things: the copper smell of blood filling the room, the wet warmth of it coating his hands, and the absolute silence where Ryan's breathing should have been.

From somewhere in the apartment, Chairman Meow released a long, mournful yowl that seemed to break whatever spell still lingered. LaMont's body began to shake, small tremors at first that built into violent convulsions. He stared at his hands, dark and wet in the moonlight, rice grains stuck to his fingers like maggots.

"No," he whispered, then louder: "*No, no, no, no . . .*"

"Ah, you're admiring the farmhouse painting, I see. Quite captivating, isn't it? The way the morning glories climb those porch posts, the children's laughter almost audible in their expressions. One can almost smell the fresh country air.

"Paintings have always fascinated me—windows to other worlds, frozen moments caught in oils and canvas. But what happens, I wonder, when the moment *isn't quite as frozen as we believe? When the window works both ways?*

"Some say art imitates life. Others suggest the relationship is more . . . reciprocal. A conversation between viewer and viewed. After all, when we gaze upon a painting, are we truly the only ones looking?

"This particular piece seems to have taken a liking to you. Notice how the light catches those children's eyes, how they seem to follow your movements? Almost as if they're waiting for an invitation.

"The price? Oh, quite reasonable for such craftsmanship. Though I should mention—it does prefer a prominent position in the home. Above a fireplace, perhaps? Somewhere it can properly . . . observe."

Bethany's sneeze echoed through the shop as dust billowed up from the cardboard box. The brass candlesticks caught the weak afternoon light, their recently polished surfaces reflecting the cluttered shelves behind her. Next to them, three porcelain figurines—shepherdesses with chipped crooks and faded pink cheeks—lay nestled in yellowed tissue paper.

Not that she needed the money from selling them. The farmhouse was paid for, inheritance taking care of that particular adult milestone. Her freelance web development work kept her comfortable, and the medical transcription she did in the dead hours between midnight and dawn padded her accounts nicely. But clearing out the attic felt like something a responsible homeowner should do. Something her grandmother would have expected.

The silence that had settled over the farmhouse in the months since Gran's funeral still felt wrong. After sixty-eight years of her grandmother's presence—her humming as she kneaded bread dough, her slippers shuffling across hardwood floors—the empty rooms now held only echoes. Katie visited occasionally with the twins, but they always left too soon, back to their busy lives in the city, leaving Bethany alone with the remnants of three generations of packrats.

"Find anything interesting up there?"

The voice made Bethany jump. The shopkeeper stood behind the counter, though she hadn't heard them approach. Their smile was pleasant enough, but their eyes were sharp and assessing, like they were cataloging every detail about her.

"Just some old things," she said, pushing the box across the counter. "My grandmother's. I thought someone might want them."

The shopkeeper lifted each item carefully, examining them in the dusty light. "Your grandmother had excellent taste. These will find good homes." They named a price that seemed generous, then paused. "But perhaps you'd like something in return? An exchange has more . . . meaning than a simple sale."

"Cash would be better," Bethany said, tucking a strand of mousy brown hair behind her ear. "I'm trying to clear things out, not collect more."

The edge of emptiness in her voice surprised her. Last night she'd eaten dinner standing at the kitchen counter again, staring at nothing while the evening news droned in the background. Just her and the creaking house.

The shopkeeper nodded, producing crisp bills from an ancient brass register. As Bethany counted them, movement caught her eye—something hanging on the far wall, partially hidden behind a massive oak armoire.

"What's that?" The words slipped out before she could stop them.

The shopkeeper followed her gaze. "Ah, the farmhouse scene? Lovely piece, isn't it? Something familiar about it."

Bethany moved closer, her shoes creaking on the warped floorboards. The painting showed a white farmhouse nestled in a green meadow, morning glories climbing its porch posts in deep purple clusters. A dirt path wound from the front steps toward the viewer. On the bottom porch step stood two children holding hands

—a girl in a pinafore dress, the boy in short pants and suspenders, their faces captured in a moment of shared laughter.

"It's like my house," she said. The resemblance was striking—the steep roof pitch, the arrangement of windows, even the morning glories matched her home's exterior. But then again, most old farmhouses around here shared similar designs, built from Sears catalogs in the '30s and '40s, before developers started carving up the land for their prefab "country estates."

"Is it?" The shopkeeper stepped beside her. "Then perhaps it belongs there. Take it."

Bethany turned. "Oh, no, I couldn't—"

"Please. I insist. Consider it a bonus for such lovely pieces." They gestured at the box of her grandmother's things. "I've gotten the better deal here, believe me."

Her grandmother's voice echoed in her head: *Nothing in life is free, child.* But the painting drew her eye—those laughing children made the farmhouse look warm, lived-in. Like her own home must have been decades before she was born, when Gran was young and the fields hadn't yet been parceled off to developers.

"If you're sure . . ." Bethany reached for the painting, suddenly wanting to see it hanging in her living room, filling the empty space above the fireplace where Gran's cross had hung until Bethany took it down last spring.

The frame was solid wood, darkened with age and heavier than she expected. Up close, the brushwork showed remarkable detail—each morning glory petal distinct, the children's faces painted with delicate care.

"Quite sure." The shopkeeper smiled. "Though you might want to hang it somewhere prominent. Above a fireplace, perhaps? Such pieces deserve to be properly displayed."

The summer heat slammed into Bethany as she left the shop. Her Subaru sat baking in the gravel lot, and she had to crack all four windows before sliding the painting onto the passenger seat. The seatbelt held it secure against the cracked leather.

The drive home took longer than usual, traffic crawling through Main Street where road crews were patching asphalt. Her AC struggled against the July humidity, and by the time she pulled into her driveway, her shirt clung to her back.

The farmhouse stood quiet against the afternoon sky. White paint peeled around the window frames—another project she needed to tackle before winter. The morning glories her grandmother had planted decades ago still climbed the porch posts, their purple blooms beginning to close in the late-day heat.

Inside, the house was still cool from running the AC all morning. She set the painting against the wall and wiped sweat from her forehead, scanning the living room. The space above the fireplace had been empty since she'd taken down her grandmother's old cross last spring. She'd meant to put something there, but like a dozen other small projects around the house, it had never seemed urgent.

The silence pressed against her ears. Her footsteps echoed as she walked through the living room, past the dining table that hadn't hosted a family meal in months. The creak of the kitchen cabinet seemed too loud as she reached for a glass of water. When had the house started feeling so hollow? She glanced back at the painting, at those children with their frozen laughter, and felt a small tug of longing. There had been a time when this house had rung with voices.

The toolbox under the kitchen sink held the drill and anchors she needed. Twenty minutes of careful measuring and cursing at the brick later, she had the painting centered and level. The children's faces brightened the room somehow, their captured laughter making the space feel less empty.

Bethany stepped back, satisfied. It looked right there, like it had been waiting for that exact spot. She checked her watch—still time to tackle more boxes in the attic before dinner. The attic still held stacks of bins she hadn't sorted, including those containing her and Katie's childhood things. Their old tea set would be up there, the delicate Limoges that Gran had let them use for "proper tea parties." Katie's teddy bear with the missing ear and her own Madeline doll were packed away too, wrapped in tissue paper alongside other fragments of a childhood that seemed impossibly distant now.

As she headed for the stairs, she could have sworn she heard children giggling, but it was probably just the neighbor's kids playing in their yard.

BETHANY RINSED her plate and loaded it into the dishwasher—one plate, one fork, one glass. The machine hummed to life, its familiar cycle filling the kitchen. She'd gotten used to eating alone in the farmhouse, but tonight the microwave pasta sat heavy in her stomach, reminding her of Sunday dinners when the table had been full.

She paused in the doorway between the kitchen and living room, looking at the painting. The evening light caught the glass differently now, making the farmhouse seem more vivid than it had in the afternoon. The children's faces appeared to be caught mid-laugh, as though responding to a joke she couldn't hear. Something about their expressions made her linger a moment longer than she'd intended.

The medical drama on TV provided background noise while she scrolled through her phone. Work emails she'd already answered, client requests she'd already scheduled. Her sister's Facebook showed the twins' seventh birthday—streamers tangled in their matching dark curls, cake frosting smeared across grinning faces. Bethany clicked away, muscle memory developed over years of seeing what she couldn't have. The doctors had been clear after her surgery at twenty-six: no children of her own, ever.

She glanced up at the painting again, at those carefree children with their perfect summer day frozen in time. Had her grandmother ever sat on those porch steps as a little girl, laughing like that? Had the house ever held that kind of simple joy for Bethany herself?

By nine, she gave up on following the TV's meandering plotlines. Upstairs, steam filled the bathroom as she stood under water hot enough to turn her skin pink. Her wet hair dampened her old cotton pajamas—the comfortable ones with faded blue stripes, not the silk ones her sister kept giving her "for when you start dating again."

The romance novel on her nightstand was pure escape

—rippling muscles, heaving bosoms, passion in exotic locations. Bethany traced the cover art with one finger, the bare-chested hero's face shadowed and mysterious. In her twenty years of dating, no man had ever looked at her the way he gazed at the heroine. Of course, she'd have to actually meet men for that to happen. The last date her sister had set up for her had been six months ago, and that had ended with awkward small talk over coffee.

Bethany set the book face-down on her chest, letting her eyes close. In her mind, the hero's shadowy face became clearer—dark eyes, strong jaw, hands that would know exactly how to touch her. Her own fingers slipped beneath the elastic of her pajama bottoms, familiar territory in the quiet dark.

The fantasy built slowly. His hands would be rough, but gentle. He'd kiss her neck, whisper that she was beautiful. That she wasn't plain, wasn't too quiet, wasn't—

THUD.

Bethany bolted upright, heart slamming against her ribs. The sound had come from downstairs, heavy and final like a body falling. Her hand flew to her chest, cheeks burning with embarrassment and fear. For a moment she couldn't move, couldn't even breathe.

Listen.

The house settled around her—familiar creaks of old wood, the cycling hum of the refrigerator, but nothing else. No footsteps, no voices. Just her pulse hammering in her throat and the lingering heat in her cheeks.

"Hello?" The word came out smaller than she intended, barely a whisper. *Stupid.* If someone was down there, she'd just announced herself. Alone. Upstairs. Her grandmother would have marched down with the old baseball bat from the hall closet, demanding to know who dared enter her home.

Bethany's phone showed 11:47 PM. Her thumb hovered over the emergency call button, muscle cramping as she gripped the device. The sound replayed in her mind—not the sharp crack of breaking wood or the hollow thump of a falling book. This had been solid, meaty.

Her legs shook as she swung them out of bed. The flashlight beam from her phone caught dust motes swirling in the air, transforming the familiar hallway into something alien. She should have a real flashlight, the kind that could actually illuminate more than two feet ahead. The kind that responsible homeowners kept in every room, along with first aid kits and emergency plans. Gran would have been prepared.

The floorboards creaked under her slippered feet— her grandmother's house announcing every movement like a betrayal. At the top of the stairs, Bethany pressed her back against the wall, straining to hear anything beyond the thunder of her own heartbeat. The darkness below seemed to pulse with each breath.

The banister felt slick under her palm, though she couldn't tell if it was sweat or just the fresh polish she'd applied last weekend. Five steps down, she stopped. The living room stretched out before her, and even in the weak phone light, she could make out the painting above the fireplace. Its surface caught the beam and reflected it back, like an eye opening in the dark.

Something lay on the white mantle beneath it.

Bethany's throat closed around a swallow. Each breath felt too loud in the silence. She forced herself forward, one step at a time, mentally cataloging potential weapons within reach. The brass umbrella stand by the door. The heavy crystal vase on the side table—her grandmother's wedding gift, never used. Her fingers tightened around her phone until the case creaked.

The shape on the mantle resolved itself as she drew closer. A crow lay twisted on the white mantle, its neck bent at an impossible angle. One wing splayed out, knocking over the small picture frame with her niece and nephew blowing out candles on their birthday cake. The bird's eyes were open, glassy and fixed, reflecting her phone's light like tiny mirrors.

"Oh God," Bethany whispered. Her stomach lurched at the sight, but she forced herself to step closer. The crow must have fallen down the chimney, panicked, and broken its neck against the painting. There was a spiderweb of lines in the aged wood just above the farmhouse's roof. But something about the bird's posture—the unnatural angle of its head, the way it had been placed directly under the painting rather than fallen in a heap of feathers—made her skin crawl. This wasn't some accident. It felt deliberate, like an offering.

Bethany lifted her light higher, examining the damage. Her breath caught. For a second, she thought the children in the painting had moved, their positions shifted. But no, there they were, exactly where she remembered, holding hands and laughing. Just her nerves making everything feel wrong, like when you wake up at 3:00 a.m. and your bedroom looks like a stranger's.

She backed away from the mantle, her mind spinning through logistics. She needed gloves. Garbage bags. Cleaning supplies for the inevitable mess. No way was she touching a dead bird with bare hands—she'd seen enough medical transcripts about infections to know better.

The kitchen drawers stuck in the humidity. She had to yank them open, the sound too loud in the quiet house. Yellow rubber gloves, usually reserved for scrubbing toilets, felt ridiculous over her trembling hands. Like playing dress-up at being brave. Gran wouldn't have hesitated—she'd dealt with dead animals on the property all her life, shooing Bethany and Katie inside whenever she dispatched an injured rabbit or bird.

Just get it done. Get it cleaned up. Go back to bed. Tomorrow it'll just be another weird story to text your sister about.

The crow's body sagged in her hands, its weight settling like guilt through the rubber gloves. Still warm— that detail made her stomach clench. Bethany fumbled with the garbage bag, doubling it with shaking fingers, trying not to look at the broken angle of its neck or those glass-bead eyes that seemed to follow her movements. Goosebumps rippled across her bare legs, and she tugged her cotton pajama top down, suddenly aware of how exposed she felt. The stairs loomed behind her, but she couldn't face them again until this was done.

The screen door creaked—a sound that was homey in daylight but felt like a scream in the darkness. July air wrapped around her, thick and alive with moisture. The chorus of crickets stuttered, then resumed, their rhythm slightly off-kilter. Her phone beam, wedged awkwardly under her arm, caught the reflective strips on the trash bins at the end of the gravel drive. They looked like eyes watching from the darkness.

Just put it in. Go back inside. Take a shower. Forget about it.

Gravel shifted under her slippers with each step. Halfway to the bins, reality slammed into her: the crow was warm. Not cooling, not room temperature—warm like something recently alive. She hadn't heard any crash during her shower, or while mindlessly watching TV. Only when she was . . .

Heat flooded her cheeks as her mind replayed the moment, her own fingers sliding beneath elastic, the fantasy shattered by that heavy thud. The garbage bag swung in her grip, its contents shifting with horrible solidity. Bethany's steps quickened, slippers scuffing against stone as she practically ran the last few yards.

The bag hit the empty bin with a hollow sound that echoed in her chest. She slammed the lid down, plastic cracking against plastic like a gunshot in the still air. The noise felt like a violation of the night's heavy silence.

Gravel crunched under her feet as she hurried back, every tree shadow reaching with grasping fingers. The back porch light threw her elongated shadow ahead of her —a dark giant that made her heart stutter when it moved. Her gloved hands slipped twice on the doorknob, leaving smears of rubber on the brass.

Inside, Bethany peeled the gloves off with shaking hands, dropping them in the sink with a wet slap. She flicked on the kitchen lights, then moved through the house turning on each lamp, pushing back the darkness.

Her throat felt tight, desperate for water, but the thought of drinking anything made her stomach roll.

She walked back to the living room, needing to check the frame was still secure. The last thing she needed was for it to crash down in the middle of the night. The overhead light cast harsh shadows above the mantle, now clean except for a few black feathers she'd missed. The crack in the frame was barely visible—just a hairline fracture in the dark wood above the farmhouse's roof. She tested the mounting—solid enough.

The painting looked different somehow. Not in any way she could articulate, but it felt like returning to a room where someone had moved everything an inch to the left. Her eyes traced over the details—the morning glories twisting up the porch posts, the distant tree line, the children's faces caught in their perpetual laughter. Nothing had changed. Nothing could have changed. Yet something had.

Bethany climbed the stairs slowly, exhaustion finally catching up with her. The adrenaline crash left her limbs heavy, her earlier arousal replaced by a hollow feeling in her stomach. Her bedroom felt different now—the romance novel face-down on her rumpled sheets, the bedside lamp casting intimate shadows—like walking in on a stranger's private moment.

She slid under the covers, letting her hand drift to where it had been before the crash. But the hero's face wouldn't form in her mind anymore. Instead, she kept seeing the crow's broken neck, those glassy eyes reflecting her light. And something else, something about the painting that nagged at the edges of her thoughts.

Bethany pulled the covers up to her chin, staring at the ceiling. The lamp cast warm shadows, making the bedroom feel safe, contained. Her heart had finally stopped racing, though her hands still trembled slightly when she thought about the crow's broken neck. She closed her eyes, willing sleep to come.

Just as she began to drift off, she heard it. Two voices this time, shushing each other between muffled laughter.

Bethany sat up slowly, her body moving before her mind could catch up. The sound was coming from downstairs, clear as crystal. Her throat went dry. The giggling stopped abruptly, replaced by the unmistakable scrape of something being dragged across hardwood floor.

Her fingers found her phone on the nightstand. The screen's blue glow showed 1:47 a.m. She swung her legs out of bed, bare feet finding the cold floor. This wasn't her imagination. This wasn't the house settling.

The scraping sound came again, followed by a soft thump. Bethany stood at the top of the stairs, phone clutched in her sweating palm. She hesitated to switch on the flashlight. Some animal instinct made her hesitate. She didn't want to announce herself.

Moonlight carved white rectangles across the living room floor through the gaps in the curtains. Another scrape, a whispered "shhh," followed by the delicate ring of china touching china.

Her toes curled against the top step, muscles cramping with the effort to stay silent. Each foot lower, the air grew colder until her skin pebbled beneath thin cotton pajamas. At the bottom of the stairs, she pressed her back against the wall, listening to the soft sounds of cups being arranged, of small things being moved with purpose.

When she finally made herself look around the corner, it was as if her lungs forgot how to work.

Moonlight poured across the living room, illuminating a scene ripped straight from her childhood. Her grandmother's Limoges tea set sat perfectly arranged on the little white table that should be gathering dust in the attic. Her old Madeline doll sat propped in the small wicker chair, facing her sister's threadbare teddy bear.

The silver spoons lay crossed exactly as they used to arrange them, waiting for their imaginary tea parties. The spoons lay crossed, one slightly crooked, because Katie had always said it was more realistic that way.

A breeze stirred the lace curtains, making shadows dance across the tiny table setting. Bethany's gaze dragged unwillingly to the painting above the mantle. Her chest constricted, each breath shorter than the last. Her mind tried to process what she was seeing. The children had moved. They'd *moved*.

The children were no longer on the porch steps in front of the farmhouse. Their faces pressed against the glass over the canvas like it was a window, so close their faces seemed to bulge from the canvas. This wasn't some trick of light or shadow—they'd physically changed position in the painting.

She couldn't look away from those eyes, couldn't stop seeing how their sweet smiles had twisted into something knowing, something patient. They watched her with hungry anticipation, small hands splayed against the canvas as if testing the membrane between their world and hers.

The nausea hit her in waves. Bethany scrambled backward until she hit the opposite wall, her fingernails scraping against the wallpaper. Her mind tried to process what she was seeing.

She had to get the painting out of the house. Right now. Nothing else mattered—not the tea set, not the impossible table, not even understanding what was happening. Get it down, get it out, burn it if she had to.

Get it out get it out GET IT OUT.

The thought pounded through her head in time with her racing heart. She couldn't spend another second with that thing in her house, couldn't bear the weight of those eyes watching her, waiting for . . . for what?

Her legs shook as she pushed herself up the wall, her sweat-slick palms leaving marks on the wallpaper. The children's eyes followed her movement. Their fingers pressed harder against the glass, creating subtle bulges in the paint that couldn't possibly be real.

Ten feet separated her from the mantle. Ten feet of moonlit hardwood between her and that thing that should never have entered her house. The tea set stood in its perfect circle, forcing her to edge around it like a trespasser in her own home. One step. Another. The porcelain cups clinked softly against their saucers though nothing had touched them.

Bethany's throat closed around a whimper. She forced herself to keep moving, to keep breathing. The painting loomed larger with each step, and now she could see the brush strokes distorting around the children's fingers, could see how the paint rippled like water around their hands. The girl's mouth opened slightly, revealing teeth too small, too numerous. The boy's eyes had gone completely black.

Don't think. Just do it. Just get it down.

Her hands trembled as she reached for the frame. The wood felt warm, alive under her fingers. Too warm. She jerked back, stumbling into the little table. Her sister's teddy bear toppled over, its glass eyes catching moonlight as it fell.

The bear hit the floor with a meaty thud—the same sound she'd heard upstairs. Bethany froze, staring at it. The worn fabric rippled, like something inside was shifting. Like something was trying to get out.

"No," Bethany whispered. "No, no, no."

But she reached for the frame again. She had to. The alternative—leaving it there, letting those eyes watch her through the night—was unthinkable.

The frame pulsed against her palms like something breathing. Bethany's fingers curled around the edges, her knuckles white. The children's faces were inches from hers now. Their eyes held a darkness that spread like ink through their pupils, bleeding into the whites. She could smell them through the glass—like old cloth and wet soil.

The painting was heavier than she remembered, fighting her as she lifted. The hanging wire caught on its nail, scraping against the wall with a sound like fingernails. Bethany's arms shook with the effort of holding it steady. One slip and she'd drop it, shatter the frame and glass, tear the canvas—

—and then what would come out?

The thought hit her like ice water. A sob caught in her throat as she finally worked it free. The children's faces contorted, their mouths opening too wide, stretching into impossible shapes. Their teeth were no longer human— row upon row of needle-sharp points filling mouths that unhinged like snakes preparing to swallow prey. Bethany stumbled backward, nearly falling as she clutched the painting to her chest, canvas facing away. She couldn't look at it anymore. Couldn't bear to see what those faces might become.

She felt movement against her chest, something scratching at the canvas from inside the frame. The wood heated beneath her fingers until she could feel her skin beginning to blister, but she didn't dare let go.

BETHANY BIT back a scream and forced herself to walk. Just walk. Through the living room, past the accusatory stares of her childhood toys. Her hip caught the corner of the tea table, sending cups rattling. Behind her, something small and soft thudded to the floor—her Madeline doll, maybe. She didn't look back.

Her hands slipped on the front door knob, once, twice, before she got it open. The night air hit her face, humid and thick. Crickets fell silent as she stumbled onto the porch. The motion-sensor light clicked on, throwing harsh shadows across the yard, across the gravel drive that led to the barn.

Twenty yards. Just twenty yards.

Bethany's bare feet slapped against wooden steps, then crunched on gravel. Sharp stones bit into her soles but she couldn't slow down. Wouldn't slow down. The painting pushed back against her arms as if trying to break free. The frame vibrated with a low hum that she felt in her teeth, in the cores of her bones.

A child's laugh rang out behind her—from the house? From the painting? She couldn't tell. Her feet moved faster, sending gravel spraying. The barn's dark shape loomed ahead, its doors hanging slightly open.

The motion light didn't reach this far. Moonlight cast everything in shades of blue and silver, turning the barn's weathered wood to corpse-gray. Bethany hit the doors with her shoulder, bursting through into darkness that smelled of dust and old hay and something deeper, earthier. Like a grave opened to the night air.

Something moved in the shadows above her. In the hayloft. Small shapes shifting in the dark. She could hear soft footfalls pattering across the wooden beams overhead, the creak of ancient boards bearing weight that shouldn't be there.

Don't look up. Don't look up. Don't look up.

Her fingers clawed at the canvas tarp she'd used yesterday to cover boxes she'd pulled down from the attic that she hadn't sorted through yet. The painting fought her grip, trying to twist free. She could feel the glass bulging, pressing outward from the ornate frame. Something scraped beneath the surface, like dozens of tiny fingernails trying to pierce through.

A giggle echoed from the hayloft. Hay rustled, shifting under small feet. Something wet dripped onto her shoulder from above—too thick to be water, too warm.

Bethany flung the tarp across a stack of boxes and slammed the painting face-down onto it. Her hands shook so badly she could barely grip the edges, but she wrapped it once, twice, three times. Something solid pushed against the layers, testing each fold. She could see shapes moving beneath the tarp, the outline of small hands pressing outward.

A soft thump behind her. The sound of bare feet on wooden floorboards.

Wrap it. Just wrap it. Don't turn around.

She fumbled with the roll of twine she'd used on the boxes, fighting to knot it around the wrapped painting. The tarp bulged and shifted under her hands. The children were moving inside, pressing against their wrappings like moths in a cocoon. Each time they pushed, the temperature around her dropped a few degrees. Her breath fogged in the suddenly frigid air.

Another thump. Closer.

The first knot held. Bethany yanked the twine tight for a second, a third. Each pull drew a muffled sound from within the tarp—not quite a cry, not quite a laugh. The painting had gone completely rigid now, as if frozen solid.

Hay drifted down from above, landing on her shoulders, in her hair. Like snow. Like ash.

Someone breathed against the back of her neck.

The breath was cold, so cold it burned. Bethany's muscles locked, her fingers still tangled in the twine. Another giggle, right beside her ear. Small fingers brushed her hair.

She threw herself backward, her hip slammed into the heavy work-table. Pain shot through her leg but she pushed off it, lurching toward the barn doors.

More laughter now, from all around. The hayloft creaked under running feet that shouldn't be there. Shadows moved in her peripheral vision—small, darting shapes that vanished when she turned her head. She caught glimpses of pale limbs, of eyes reflecting moonlight like animals', of mouths stretched too wide.

The motion light had gone dark. Twenty yards of moonlit gravel stretched between her and the house. Bethany's feet hit the barn threshold and she ran. Gravel bit into her soles but she couldn't feel it, couldn't feel anything but the cold spot on her neck where something had breathed.

Behind her, the barn doors banged shut. The sound echoed across the empty field like a gunshot.

Don't look back don't look back don't look—

She hit the porch steps running, almost falling. The front door stood open. Inside, she slammed the door and threw the deadbolt, then slid down against it, gulping air.

In the sudden quiet, she heard it. A sound that turned her blood to ice. The soft clink of porcelain against porcelain.

Bethany turned, still pressed against the door. The tea cups rattled in their saucers, spinning slowly clockwise. Her sister's teddy bear sat upright again, its head tilted at an impossible angle. The Madeline doll stood beside it— *stood*, on legs that were supposed to be cloth-stuffed and limp.

Dark liquid began to pour itself into each cup, defying gravity, flowing upward from the floor. It didn't splash or overflow, just coiled and twisted in the porcelain cups, never still. It looked like ink, but thicker, with a metallic sheen where moonlight caught its surface.

The children from the painting stood on either side of the table. Their faces sharp, too dark, like photographs left too long in the sun. Paint dripped from their clothes, hitting the floor with wet slaps before crawling back up their legs. Their skin had a waxy, unfinished quality—like dolls crafted by someone who'd only seen humans in poor light.

The girl's head twisted toward Bethany, neck bending at angles that made Bethany's vision swim. The girl's smile split her face like a knife wound, teeth too numerous, too sharp. When she spoke, her voice sounded muffled, as if something were caught in her throat.

"Join our tea party."

The boy's hands elongated, fingers stretching across the room like pulled taffy, leaving trails of wet paint in the air. They wrapped around Bethany's wrist, cold and slick and strong. Where they touched, her skin began to ripple, colors bleeding up her arm like watercolors on wet paper.

Bethany tried to jerk away but the boy's fingers sank into her flesh like hooks. The cold spread up her arm, numbing everything it touched. She thrashed, kicking out, but her foot passed through the girl's leg with a wet, horrible sound.

The room tilted sideways. Bethany's head spun as gravity shifted, the ceiling becoming the floor, then the wall, then nothing she could understand. Her childhood toys convulsed, twisting into shapes her mind refused to process. The teddy bear's glass eyes watched her, and watched her, and watched her—too many eyes now, multiplying like cancer.

The girl moved closer yet, in that broken-film way, each step existing in the wrong order. Bethany could hear her bones clicking, could hear the sounds of her joints moving in ways bones were never meant to move.

The boy's other hand shot out, grabbing Bethany's free arm. His touch spread numbness everywhere it landed. Her fingers tingled, then went dead. She couldn't feel her own pulse anymore. Bile rose in her throat as the numbness crept past her elbow, stealing every sensation except the thundering of her heart.

The room spun again, and Bethany threw herself sideways, slamming her shoulder into the wall. The impact sent lightning through her chest, but the pain cut through the numbness. She sucked in a ragged breath and wrenched herself free of the boy's grip. The feeling rushed back in a wave of needles.

She lurched toward the hallway, her legs shaking so

badly she could barely stand. The children's feet pattered across the hardwood behind her—too many footsteps, coming too fast, a staccato rhythm that didn't match any possible number of feet.

Bethany hit the hall running, her heart hammering. The darkness ahead stretched forever, the staircase at the end impossibly distant. Each gasping breath burned in her chest. The hall light flickered once, twice, then died.

A soft scraping sound came from above.

Bethany's gaze jerked up, though every instinct screamed not to look. The girl crawled across the ceiling. Her head twisted completely backward to maintain eye contact, and when she smiled, her mouth kept spreading, wider and wider, until it bisected her entire face.

Bethany lunged for her bedroom door, but it slammed shut before she could reach it. The wood rippled like water, and the boy's face pressed through, features distorting as he emerged. His fingers splayed across the door's surface, each one stretching longer, reaching for her hair. She could hear the wood creaking, splitting around his body.

Her legs nearly gave out. The numbness in her arm had spread to her shoulder, turning her skin to ice. Only the staircase remained an option, a black mouth leading up.

Behind her, the girl's joints popped and crackled as

she descended from the ceiling. Bethany could hear her vertebrae snapping into new configurations, could hear the wet slide of tendons rearranging themselves. The sound was so close now—right behind her ear, where her neck was most vulnerable.

Bethany bolted for the stairs. Her bare feet slapped against wood, then carpet. Three steps up, something grabbed her ankle. The touch was like plunging her leg into frozen water. She kicked out blindly and connected with something soft that yielded like rotten fruit. The grip vanished.

Six steps. Eight. Her lungs burned. The stairwell stretched above her, the landing seeming to retreat with each step. From below came the sound of small feet, moving in that arrhythmic pattern that made her brain stutter.

Ten steps up. Eleven. Bethany's vision blurred with each heartbeat.

A laugh echoed from above—high and clear. Bethany's head snapped up. Through tears and terror she saw the boy standing on the landing, his silhouette against the darkness. His neck stretched sideways, head lolling against the wall while his body faced her. The sound came again, and she realized it wasn't laughter at all.

It was her grandmother's voice.

"Bethany, dear. Come upstairs. I've been waiting for you."

Her foot missed the next step. She caught herself on the banister, but the wood was soft under her fingers, pulsing like something alive. Bethany jerked her hand back and stumbled upward, using the wall instead. Behind her, the girl's footsteps grew closer, that impossible rhythm making Bethany's teeth vibrate in her skull.

The landing was three steps away. Two. The boy hadn't moved, but his shadow had. It stretched across the floor toward her, and where it touched the carpet, the fibers twisted themselves into tiny, grasping fingers.

Something cold pressed against the back of her neck—a small hand, fingers splayed. Bethany's spine turned to ice. She threw herself forward, crawling up the last step on hands and knees. Her palms left bloody prints on the carpet. When had they started bleeding?

"There you are, dear. I've made your favorite cookies." Her grandmother's voice came from the boy's mouth, but wrong—like words played backward then forward again. His face remained a child's, but his eyes held ancient recognition. "Don't you want to give your Gran a hug?"

Bethany scrambled backward on the landing, leaving scarlet smears across the carpet. The boy's neck kept uncoiling, his head weaving like a snake's. His mouth gaped wider until his face split vertically, revealing row after row of teeth in the darkness of his throat. From inside came her grandmother's voice, perfectly clear:

"You've let the house go, Bethany. I'm very disappointed in you."

The temperature plunged. Bethany's breath came out in white clouds. Frost crackled across the baseboards, across the family photos on the wall. Behind their glass, the faces in the pictures turned to watch her—her parents at their wedding, Katie's graduation, even the portrait of Gran that usually hung in the dining room. All watching with black, empty eyes.

The girl's footsteps stopped.

Bethany forced herself to look back at the stairs. Nothing. Just darkness and the bloody prints she'd left. But the silence pressed against her eardrums like a physical weight. Even her ragged breathing seemed muffled, distant.

A soft creak came from her right—the attic door swinging open on its own. Cold air poured down the narrow stairs, carrying with it the smell of dust and rot and something older, something that reminded her of the time she'd found a dead fox beneath the porch, its body crawling with maggots.

Movement caught her eye. The girl stood at the bottom of the main stairs now, head cocked at an angle that made Bethany's neck ache just looking at it. She took one step up. Another. With each step, her legs lengthened, joints crackling as they reformed.

The boy's shadow crept across the landing toward Bethany's legs. She could feel its cold even before it touched her. The attic stairs creaked again behind her.

They were herding her up.

And she couldn't feel her legs anymore.

BETHANY TRIED to stand but her legs buckled, sending her sprawling. The carpet scraped her palms raw. She could feel the boy's shadow sliding up her calves, leaving trails of numbness in its wake. Her body felt wrong, disconnected, like she was trying to control a puppet with frozen strings.

The girl reached the middle of the stairs, her neck elongating with each step. Bethany could see the individual vertebrae pressing against skin that had gone waxy, translucent. The bones shifted and clicked as the girl's head swayed from side to side, tracking Bethany's movements.

A sound drifted down from the attic—soft at first,

then clearer. The scritch-scratch of something moving across wooden floors. Bethany's stomach clenched. More children. How many were there?

She dragged herself toward the wall, leaving crimson streaks behind. Her fingers found the baseboard, then the doorframe. The attic stairs stretched up into darkness, each step visible only by the frost that outlined its edge.

Bethany pulled herself up using the doorframe, her arms shaking with the effort. The scratching sounds grew louder. Multiple sets of feet shifted in the darkness above, waiting.

"No one wants you," her father's voice now, layered over her mother's. "We left because of you."

The girl's hand shot out, fingers wrapping around Bethany's ankle. This time the grip held.

Bethany kicked wildly, but her numbed leg barely responded. The girl's fingers sank deeper, pressing between tendons until they touched bone. Cold radiated up Bethany's calf, turning her muscles to stone. She clawed at the doorframe, fingernails splitting against wood, but she couldn't stop herself from being dragged backward.

"Please," she gasped, "please don't—"

The girl's other hand seized Bethany's free ankle. One sharp yank pulled her legs out from under her. Her chin cracked against the first attic step, filling her mouth with blood. The metallic taste mixed with the dust, making her gag.

They dragged her up the narrow attic stairs, step by painful step. Bethany fought with everything she had, but the numbness spread through her body with each passing second. She caught glimpses of the framed photos on the wall. Every face was distorted, eyes following her progress up the stairs.

More hands reached down from above—small fingers stretching from the darkness, grasping at her hair, her shoulders, her arms. Each touch spread that horrible numbness deeper. Bethany thrashed, but her body felt distant now, like she was sinking into ice water.

"Get off me!" she screamed, her voice breaking. "Let me go!"

The attic air hit her face like a slap of winter wind. Her body scraped across the threshold, over rough wooden boards that tore at her skin. Moonlight filtered through the single window, casting knife-blade shadows across the floor.

Children stood in every corner, between boxes and old furniture. Dozens of them. Their faces all bore the same waxy pallor, the same hungry eyes. Some appeared more solid than others—the newer ones still dripping paint that pooled at their feet, while others seemed almost transparent, faded photographs of children who had been trapped for decades, perhaps centuries.

And there, propped against the wall directly in front of her, was the painting she'd left wrapped in the barn.

The tarp lay in shreds around the painting's frame. The glass surface now rippled like the surface of a pond, more faces pressing through from the other side, trying to emerge.

The hands released her suddenly. Bethany collapsed onto the attic floor, her body a dead weight. Cold boards pressed against her cheek. She tried to move, to crawl toward the stairs, but her limbs wouldn't respond.

The boy and girl from the painting stepped into her field of vision. Their faces were different now—the malice replaced by something like longing. The girl reached out, her cold fingers brushing Bethany's cheek in an almost tender gesture.

"Mother," she whispered, the word echoing as the other children took it up. "Mother. Mother. Mother."

The word stabbed through Bethany like a knife. The one thing she could never be. The doctor's words after her surgery at twenty-six echoed in her mind: "I'm sorry, but you'll never be able to have children."

"No," Bethany croaked, finding her voice. "I'm not— I can't—"

"You'll be our mother now," the boy said, his voice carrying the same unnatural echo. "Forever."

The children formed a circle around her, their movements synchronized, hands extended. Bethany fought against the paralysis, managed to push herself up onto her elbows.

"I won't," she gasped. "I'm not your mother. I'm not anyone's mother."

The girl's face crumpled, an expression of raw hurt that tugged at something deep inside Bethany despite her terror. "Everyone leaves us," she said. "They always leave."

"We need a mother," the boy continued. "Someone to stay with us. Forever."

The children began to hum, a sound that vibrated through Bethany's bones. The painting behind them pulsed with each note, its surface stretching outward like a membrane about to tear.

Bethany managed to get to her knees, her muscles screaming in protest. "I'm going to leave now," she said, trying to keep her voice steady. "I'm going downstairs, and I'm never coming back up here."

The girl's face hardened. "No one leaves."

The humming grew louder, and Bethany felt her body being pulled toward the painting. The air itself seemed to flow toward it, carrying dust motes and her breath in an unnatural current. She dug her fingers into the floorboards, feeling splinters drive deep under her nails.

"No!" she screamed, her voice raw. "I won't go! I'm not your mother!"

The boy's expression remained unchanged as he stepped forward. "You will be."

He grabbed her wrist, his touch burning cold. The girl took her other arm. Together, they pulled her toward the painting with shocking strength. Bethany's knees dragged across the floor, leaving bloody trails as she fought against them.

"Please," she sobbed, "please don't do this—I don't want this—"

The other children joined in, small hands grabbing at her clothes, her hair, her legs. She was lifted from the ground, suspended in the air by dozens of small hands as they carried her toward the painting. The glass surface bulged outward, reaching for her.

Bethany twisted violently in their grip, managing to wrench one arm free. She clawed at the nearest face—the girl's—feeling her fingers sink into something with the consistency of wet clay. The girl didn't flinch, didn't even blink as Bethany's fingers dug furrows in her cheek.

Then Bethany's back hit the painting's surface. But instead of solid glass, she felt herself sinking into it, like falling backward into freezing water. The children pushed her deeper, their small hands surprisingly strong. She could feel the surface tension of the painting closing around her, cold and viscous.

"No!" she screamed one last time, her voice distorted as the painting began to swallow her. "I'm not your mother! I don't want to be your mother!"

The girl leaned close, her face inches from Bethany's. "But we chose you," she whispered.

The last thing Bethany saw was the boy and girl smiling down at her, their perfect, painted faces restored as she sank deeper into the canvas. Cold colors washed over her, transforming her flesh, remaking her into something else. She felt her joints stiffening, her skin hardening into brushstrokes.

Her scream died in her throat as the painting sealed itself around her.

THREE DAYS LATER, the shopkeeper carefully hung the painting in their window display. The farmhouse remained unchanged, the morning glories still climbing its porch posts. But now the two children on the steps had company—a woman stood behind them, her hands resting on their shoulders in a maternal gesture.

To most observers, the woman's face might appear serene, motherly. But if one looked closely, one might notice the terror frozen in her eyes, the slight parting of her lips as if caught mid-scream.

The shopkeeper stepped back, admiring the effect. The painting would find a new home soon enough. Perhaps with that nice young couple who'd been browsing last week. The ones who'd mentioned how quiet their house was, how they hoped to fill it with children someday.

The bell above the door jingled as a customer entered.

"Ah," the shopkeeper said, turning with a pleasant smile. "Welcome to our little charity shop. See anything that catches your eye?"

"Ah, welcome. I see you've noticed the leather jacket in the window display. Quite striking, isn't it?

"This particular jacket once belonged to a man named James Kessler. Perhaps you've heard of him? No? A shame how quickly society forgets its monsters.

"James had a particular talent for violence, you see. Seven young women fell victim to his attentions before the state finally strapped him to a gurney in 1989.

"A couple of decades later, young Elliot Landry—painfully invisible to his colleagues, especially the woman he adored from afar—found himself drawn to this jacket, just as you are now. His transformation was remarkable. Suddenly seen, suddenly confident.

"A jacket like this offers many things—strength when you're weak, courage when you're afraid, words when you're silent. The question becomes: how much of yourself are you willing to surrender for these gifts? And once surrendered, can you ever truly get it back?

"But confidence, like power, comes with costs. The line between wearing a jacket and being worn by it can blur so easily. Perhaps you'd like to try it on? The price is quite reasonable, I assure you."

PROLOGUE—DEATH ROW, 1989

The leather jacket creaked softly as James Kessler walked down the long corridor, each step measured and unhurried. Unlike most men making this final journey, he didn't shuffle or stumble. His posture remained perfect, shoulders squared beneath the worn leather bomber jacket—just like the one Indiana Jones wore—that had become his signature.

"Last chance, Kessler," the guard to his left muttered. "Want to tell us where the others are buried?"

James smiled, the expression reaching his eyes in a way that made the guard look away. The jacket shifted with his movements, almost alive. Seven years it had been his companion, through all his special moments. Through all their deaths.

"Now, Tommy," he said, noting with satisfaction how the guard flinched at the use of his first name, "you know I prefer to keep some mysteries. Gives people something to talk about after I'm gone."

The execution chamber door loomed ahead, sterile steel reflecting the fluorescent lights. James could see his reflection in it—still handsome at forty-three, not a gray hair in sight. The jacket's collar brushed his neck like a lover's caress.

"Besides," he continued as they reached the door, "someone worthy might want to follow in my footsteps someday. Can't make it too easy for them."

The guards exchanged glances. Even after seven months on death row, James Kessler unsettled them. Most killers raged or broke down. He just watched, smiled, and wore that jacket like armor.

The chamber smelled of disinfectant and fear. James breathed it in like perfume as they strapped him to the gurney, the leather jacket bunching uncomfortably under his back.

"The jacket needs to come off, Kessler," the older guard said, not meeting his eyes.

"Now that," James replied, his voice soft but carrying to every corner of the room, "is going to be a problem." He flexed against the restraints, making the leather creak. The sound echoed off the tile walls, too loud in the small space. "The jacket stays."

Through the observation window, he could see the witnesses gathering. Families of his victims, he supposed, though he had trouble remembering their faces. The girls, yes—he remembered every detail of his special ones. But their families? They were background noise, static in the symphony of his work.

"Protocol says—" Tommy started, but James cut him off with a laugh that made everyone in the room take a step back.

"Protocol?" The word dripped with amusement. "Tommy, I'm about to die. What exactly do you think you're going to do if I refuse? Kill me?"

The prison chaplain stepped forward, Bible clutched to his chest like a shield. "Would you like to make your peace with God, my son?"

James turned his head just enough to fix the chaplain with his gaze. The man's Adam's apple bobbed as he swallowed. "God?" James said, smiling. "God's got nothing to do with this." His eyes drifted to the observation window, to his distorted reflection in the glass.

The technician struggled to push the leather sleeve of the jacket up high enough to find a vein, then the IV line slid into his arm with practiced efficiency, the tech's hands steady despite James's unwavering stare. The leather of his jacket creaked again as he settled back, the sound almost drowning out the soft prayers of the chaplain.

"Any last words?" the warden asked, his voice carefully neutral.

James looked at his reflection in the observation window one last time. The jacket fit him perfectly, as it had since the day he'd found it. Or had it found him? He could never quite remember anymore. The leather seemed to pulse against his skin, alive with possibilities.

"This isn't the last you've seen of old Jimmy Kessler," he said, his voice carrying clearly through the chamber. "I'll be back soon enough and God help you all."

The first chemical entered his bloodstream, and James felt a curious warmth spreading through his limbs. Similar to the warmth he'd felt when he first put on the jacket, when it had whispered to him of all the wonderful things they could do together. His vision began to blur, but he kept his eyes fixed on his reflection, watching as the leather seemed to ripple and shift.

The second chemical hit. James's heart stuttered, but his smile never wavered. In the window's reflection, he thought he saw something move within the jacket's shadows, something darker than leather, older than death.

The final injection began. As consciousness faded, James Kessler whispered words that only the jacket heard: "Find someone worthy. Someone who needs us. Someone who . . ."

His voice trailed off. The heart monitor flatlined. But in the reflection, for just a moment, his smile seemed to linger after his face had gone slack.

The jacket was removed and logged into evidence, still warm despite the chamber's chill. By morning, it would be gone, leaving only an empty evidence bag and a confused clerk who could never quite remember what had happened to it.

PRESENT DAY

THE RAIN HAD SOAKED through Elliot's sport coat by the time he was halfway to the office. He'd missed the bus by just a few minutes and he was already running late to the stupid office party he didn't even want to go to.

Water trickled down his neck as he waited at a crosswalk, his dress shoes already squeaking. The few other pedestrians hurried past with umbrellas and purpose, not sparing him a glance. Typical of his life—even in a crowd, he remained invisible.

His phone buzzed and he pulled it from his pocket, doing his best to protect it from the rain with his free hand cupped over the screen. *Where are you? Brad's being Brad. Could use a friendly face.*

Elliot's heart jumped, then immediately sank. Friendly face. That's all he'd ever be to Maggie. The safe one. The nice one. The one who helped her with Excel formulas and listened to her complain about Brad's latest thoughtless comments or infidelities. Never the one she actually wanted.

The light changed, and Elliot stepped into the crosswalk. Through the curtain of rain, the storefronts glowed like distant planets—a coffee shop, a closed bookstore, a narrow shop he'd never noticed before. Something in its window caught the streetlight, drawing his eye.

He stopped, letting other pedestrians flow around him like water. In the window, a leather jacket hung on a simple wooden hanger. Not some flashy modern thing, but worn leather the color of dark honey, creased with character. The kind of jacket that suggested stories, adventures. The kind of jacket meant for a rugged explorer. The kind of man he was not. Yet he felt a strong and immediate attraction to it and he found himself entering the shop.

The door opened silently when he pushed it, a small bell offering a muted chime. Warm air washed over him, carrying a strange mix of scents, leather and dust and something else, something that made him think of old photographs and forgotten memories.

"Ah," said a voice from the shadows. "I was wondering who would notice it tonight."

"Excuse me?"

"The jacket. It's what brought you in, is it not?"

The shopkeeper emerged from between crowded shelves, their features oddly difficult to focus on in the shop's amber lighting. They moved with fluid grace, stopping before the jacket's display.

"Yes," Elliot said, pushing his rain-spotted glasses up his nose. Water dripped from his hair onto the worn wooden floor, each drop seeming to disappear the moment it landed. "It's not my style, of course. Not sure I could even pull it off."

"No?" The shopkeeper's smile was a curious thing, more suggested than seen. "And what exactly is your style, Mr. . . . ?"

"Landry. Elliot Landry." He stepped closer to the jacket, drawn by the way the leather caught the light. "I guess I don't really have one. Just . . . dull and uninspiring, I suppose."

"Ah, but dull and uninspiring are just masks we wear, aren't they?" The shopkeeper lifted the jacket from its display with surprising reverence. "Sometimes what we need . . ." They held out the jacket. ". . . is to try on something new, Mr. Landry."

Elliot reached out hesitantly, his fingers brushing the leather. A curious warmth spread up his arm, as if the jacket had been sitting in sunlight instead of a darkened window. "It looks expensive."

"Oh, don't be so sure. This is a place of second chances and the prices reflect as such." The shopkeeper's smile deepened. "Try it on. See how it fits."

The leather was butter-soft as Elliot slipped his arms into the sleeves. It should have been too big or too small —clothes never fit him quite right—but the jacket settled onto his shoulders like it had been made for him.

"Perfect," the shopkeeper murmured. "As I knew it would be."

Elliot turned toward the shop's clouded full-length mirror. His reflection straightened to its full height—a good two inches taller than his actual height, his shoulders appearing much broader than they were. He blinked, startled by the way a quality piece of clothing could make him appear more formidable and confident. The reflection blinked back, but held itself with an assurance he'd never managed before.

"How much?" he heard himself ask.

The shopkeeper named a price so reasonable it should have been suspicious. But Elliot was already pulling out his wallet, unable to look away from how the jacket transformed his reflection.

One simple piece of clothing, and suddenly he looked like someone who belonged in board rooms instead of back offices. Someone who commanded attention rather than avoided it.

"An excellent choice," the shopkeeper said as they dropped Elliot's wet jacket into a brown paper bag. "Or perhaps it chose you?"

THE RAIN HAD STOPPED by the time Elliot left the shop, though he couldn't remember exactly when. His old jacket sat wrapped in brown paper bag under his arm, all but forgotten. The leather one fit so well, as if it were tailored specifically for him, and he found himself standing straighter, his steps more purposeful.

The office building where he spent eight to ten hours a day loomed ahead, light and music spilling from the third-floor windows. Usually, the sight of this place would make his stomach clench with anxiety. It wasn't the work, because he loved what he did. But it was the people, the politics, the drama, the cliques. It was like he was still in high school, still fodder for bullies.

Fuck that, kid. You're gonna walk in there like you own that room.

Elliot stopped dead in his tracks, his heart slamming against his ribs. That voice—rough, aggressive, alien— had felt like sandpaper scraping across his thoughts. He jerked his head around, half-expecting to find someone standing behind him, but the street was empty.

His hands trembled as he glanced at his reflection in the lobby window. For a split second, he could have sworn someone else looked back—a stronger jaw, different eyes. He stumbled backward, pulse roaring in his ears. When he blinked, it was just him again.

The elevator ride up felt shorter than usual. The jacket's leather creaked softly as he adjusted his collar, the sound oddly comforting. Like a friend whispering encouragement.

Music and voices spilled into the hallway as he

approached the office party. He reached for his glasses to adjust them—a nervous habit—but stopped halfway. Instead, he ran a hand through his rain-damp hair, mussing it slightly. It felt right somehow.

"Elliot?"

He turned. Maggie stood in the hallway, a plastic cup of wine in her hand. Her eyes widened slightly as she took him in.

"Hey," she said, head tilting. "Is that a new jacket?"

"Just picked it up," he heard himself say, his voice carrying a confidence he didn't recognize. "Like it?"

She stepped closer, free hand reaching out to touch the leather sleeve. "I do. It's not your usual style but it looks really good on you."

She's noticing you, kid, that voice whispered in his head. *If you let me, I'll show you how to make her notice more.*

Elliot's pulse quickened. It was that same rough voice he'd heard outside.

Relax, kid. You felt it earlier when you slid on the jacket, how good it was to stand tall, to be seen, the voice responded with a warmth that somehow cut through his rising panic. *Just relax. Here, let me help you.*

Elliot leaned against the hallway wall, a casual movement that felt foreign yet natural. "Brad inside?" The question came out with just the right note of disinterest, though he couldn't remember deciding to ask it.

"Being his usual charming self," Maggie rolled her eyes, taking another sip of wine. "He's telling that stupid college football story again. You know, the one where he supposedly won the big game?"

She's uncomfortable with Brad. Use that, kid. The voice again, smoother now. *Watch her body language. The way she keeps looking at the door like she's avoiding something. Or someone.*

"Sounds thrilling," Elliot said, surprising himself with the dry sarcasm in his tone. "Want to get some air?"

The words had barely left his mouth when panic seized him. He'd never be so forward with Maggie—their whole friendship balanced on him being her safe space, her sympathetic ear. But Maggie was already smiling, relief flooding her features.

"God, yes," she said, then glanced down at her empty cup. "Let me just grab another drink first?"

As she disappeared into the party, Elliot caught his reflection in the chrome elevator doors.

"Smooth, kid," his reflection said. "Way to separate her from the herd."

Elliot jerked back. The voice had been clear that time, not in his head, but from the reflection itself. His mirror image smiled, the expression nothing like his usual uncertain grin. It was confident, predatory even.

"Who—" he started to ask, but Maggie was returning, two fresh drinks in hand.

"One for you," she said, offering a cup. "You look like you saw a ghost. You okay?"

Take the drink and smile, kid. We'll talk later and this will all make sense. Just go with it.

"I'm fine. Shall we?" He gestured toward the small balcony where smokers usually gathered. It was empty now, everyone inside avoiding the damp night air.

As he followed Maggie out, he caught one last glimpse of his reflection. It winked at him.

The night air carried the lingering chill of rain, and Elliot instinctively reached for his jacket zipper. *Leave it on,* James suggested. *She's made it through plenty of cold nights without your help, kid.*

Maggie rested her elbows on the balcony railing, her dark hair falling forward as she looked down at the city below. The same way she always perched on the edge of his desk during lunch breaks, venting about deadlines or office politics or Brad. Tonight though, something felt different. He wasn't just her safe harbor, her reliable shoulder to cry on.

"Remember that client presentation I was freaking out

about yesterday?" she asked, turning to face him. "The one I thought I'd totally bomb?"

Don't jump in with reassurance this time, James advised. *Let her finish.*

"The quarterly projections?" Elliot leaned against the railing beside her, close enough to catch the scent of her perfume but not so close that she'd pull away.

"I crushed it. Absolutely crushed it." Her smile lit up her whole face. "Though I probably drove you crazy stress-texting you all morning about it."

"You did good work. I knew you would." The words came out warm but measured, not his usual eager praise.

Maggie tilted her head, studying him with curious eyes. "You seem different tonight. More . . . I don't know. Present?"

Before he could respond, shouting erupted from inside —Brad's voice carrying over the party noise. Maggie's shoulders tensed, that familiar mix of embarrassment and resignation crossing her face. The same look she'd worn last month when Brad had "forgotten" about her birthday dinner.

"Sounds like Brad's making friends again," Elliot said, the dry observation coming easier than his usual careful diplomacy. He watched her twist her ring—a nervous habit he'd noticed the day Brad gave it to her.

"He's just . . . celebrating. Big sale today." Maggie's

defense lacked its usual conviction. She glanced at Elliot, then quickly away, like she always did when the conversation veered too close to Brad's behavior. Too close to the understanding that hung between them.

She knows you see through it, James whispered.

Through the glass doors, he caught another glimpse of his reflection. The other man smiled back, handsome and dangerous in a way Elliot had never been. The jacket's leather seemed to ripple in the dim light, like something moving beneath its surface.

"I should go back inside," Maggie said suddenly, but she made no move to leave. "Brad will wonder where I am."

Let him wonder, the voice whispered, and Elliot found himself repeating the words aloud, adding, "You deserve better than wondering."

Maggie turned back to him, something shifting in her expression. "Elliot . . ." she started, but whatever she meant to say was cut off by the balcony door slamming open.

Brad filled the doorway, his broad shoulders blocking the party lights. "There you are, babe. Been looking everywhere." His words slurred slightly, eyes narrowing as they fixed on Elliot. "Hey, Laundry. Nice jacket. Raid your grandfather's closet?"

Don't react, kid. Let this asshole dig his own grave.

Elliot stayed silent, letting his smile speak for him.

"What are you two doing out here?" Brad asked, his eyes bouncing back and forth between them. "You wouldn't be out here trying to hit on my girl, now would you, Laundry?"

"Brad, stop," Maggie said, "Elliot was just—"

"Being a good listener?" Brad's laugh was just shy of cruel. "That's our Elliot, right? Always there to hear about your problems. Like your own little human diary." He reached for Maggie's arm. "Come on, babe. We're here to party so let's get back to it."

You're not gonna let that asshat put his hands on her, are you, kid?

"Maybe she wants to stay," Elliot said, the words coming out in a voice that wasn't quite his own.

Brad's head snapped around, genuine surprise crossing his features. "What did you say, Laundry?"

Don't back down now. Show her who you really are.

"I said," Elliot stepped forward, the jacket creaking softly, "maybe you should ask what Maggie wants."

The air seemed to thicken with tension. Brad released Maggie's arm, turning to face Elliot fully. "Since when do you have an opinion, dickwad?"

Since me, the voice purred, and Elliot felt something dark and warm unfurl in his chest.

"Guys, please, let's not do this," Maggie said, concern in her eyes.

The fluorescent light from the party cast harsh shadows across Brad's face as he stepped closer. He had four inches and forty pounds on Elliot, and his smile suggested he was very aware of this fact.

"You know what your problem is, Laundry?" Brad's breath reeked of expensive scotch and mints. "You don't know your place."

Oh, but we do, the voice whispered. *Show him.*

The jacket tightened against Elliot's skin, sending waves of warmth through his body. The fear that began to creep over him when Brad squared up evaporated like rain on hot pavement, replaced by something darker, something hungry. He found himself smiling up at Brad, not backing away.

"And where exactly is my place, Brad?" The words came out soft, almost intimate. "Cowering in a corner somewhere, watching while you treat Maggie like she's your property?"

"Guys, stop," Maggie stepped between them, her hand pressing against Brad's chest. "You're drunk, Brad. Let's just—"

"No, no," Brad brushed her aside. "I want to hear what else Laundry has to say. Come on, tough guy. What do you want to say to me?"

He's going to try to hit you. When he does . . .

The voice trailed off as Brad's fist swung toward Elliot's face. Time seemed to slow. The jacket tightened, moved, guided his body like a puppet master with familiar strings. Elliot stepped inside Brad's swing, smooth as silk, and suddenly he was behind the larger man, one hand gripping Brad's extended arm.

"What the f—" Brad's surprise turned to pain as Elliot applied pressure to his wrist. Just the right spot, like he'd done it a hundred times before.

Like WE'VE done it a hundred times before, kid, the voice corrected with satisfaction.

"Elliot!" Maggie's voice seemed to come from far away. "Let him go!"

The jacket's warmth had become a burning heat, and Elliot could feel something else moving beneath his skin, trying to take control. In the balcony door's reflection, he caught a glimpse of another face overlaying his own, handsome, cruel, smiling with too many teeth.

Make him hurt, kid, the voice urged. *Make it clear he can't mess with you anymore.*

Brad's wrist made a soft popping sound as Elliot increased the pressure. Not enough to break, just enough to promise what could happen. Brad's face had lost its flush of alcohol, turning pale beneath the fluorescent lights.

"What the hell, Landry?" he gasped, dropping the silly nickname, trying to pull away.

"You should be more careful, Brad." The words came from Elliot's mouth, but the voice wasn't his own. Deeper, smoother, filled with dark promise. "You might just mess with the wrong person."

"Elliot, please." Maggie's hand touched his shoulder, and the contact broke whatever spell had taken hold. Elliot released Brad's wrist, stepping back. The jacket's heat receded, leaving him cold and confused. What had just happened? How had he known that hold?

Brad cradled his wrist, all pretense of superiority gone. His eyes held something new when they looked at Elliot—fear.

"You're fucking crazy," Brad spat, but he was already backing toward the door. "Come on, Maggie."

Maggie stood frozen between them, her fingers pressed to her lips. The city lights caught the tears threatening at the corners of her eyes, and something twisted in Elliot's chest. This wasn't how tonight was supposed to go.

"I should go," she whispered, but her feet didn't move. The space between them felt charged with everything they'd never said, everything they couldn't say. "I'm so sorry—"

"You don't need to apologize for him." The words

came out soft, heavy with years of watching her make excuses for Brad, for everyone.

She met his eyes then, and for a heartbeat, the world narrowed to just them. Her lips parted like she might say more, but Brad's voice shattered the moment: "Maggie! Now!"

She took a step backward, then another, her eyes never leaving his face. When she finally turned away, it felt like a door closing between them.

The party swallowed them up, leaving Elliot alone on the balcony. His hands wouldn't stop shaking from adrenaline, from loss, from whatever power had surged through him moments ago. He pressed his palms against the cold railing, trying to ground himself in something real.

Movement caught his eye. In the glass door's reflection, he saw the other man clearly now. Older, devastatingly handsome, wearing the jacket like he'd been born in it. The reflection winked at him.

Well done, it mouthed.

ELLIOT'S HANDS shook as he unlocked his apartment door. The night felt surreal—that surge of power with Brad, the charged moment with Maggie on the balcony, all those unspoken things hanging between them. Now, in the silence of his empty apartment, doubt crept back in. He'd crossed a line tonight, shattered the careful balance they'd maintained for so long.

He flicked on the bathroom light, bracing himself before looking in the mirror. His reflection stared back, then it changed. The other face overlaid his own, the older, handsome man, wearing the jacket like it had been made for him.

"Who are you?" The words escaped as barely a whisper. Elliot's reflection rippled, like heat waves off summer pavement, and another face emerged through his own. His heart slammed against his ribs as reality twisted inside his bathroom mirror.

Elliot's legs threatened to give out. He gripped the bathroom counter, knuckles white against the porcelain. This couldn't be real. People had breakdowns all the time, didn't they? Stress, trauma, guilt—any of those could cause hallucinations. He should call someone. His sister maybe. Or 911. Tell them he was seeing things, hearing voices. They'd help him. They'd make this stop.

"Someone who can help you get what you want," the

stranger said, his voice warm and persuasive. "Or should I say who you want?"

The bathroom light flickered, casting strange shadows across both their faces. Elliot watched, horrified and fascinated, as his own movements synced with the stranger's—a slight tilt of the head, fingers drumming against the counter. Like a puppet and its master, but he couldn't tell which was which.

"This isn't—I need to sit down." But Elliot couldn't look away from the mirror, couldn't break the strange connection between his movements and the stranger's. The small bathroom felt like it was closing in, the walls pulsing with each frantic beat of his heart.

"Feeling overwhelmed?" The stranger's smile was oddly gentle. "That's natural. Most people would have run screaming by now. But you're still here, aren't you? Still wondering what else is possible?"

He was right. Terror clawed at Elliot's throat, but beneath it lurked something else— a desperate curiosity. The strength he'd felt earlier, that surge of confidence that had let him stand up to Brad . . . had that been real? Or was he truly losing his mind?

"What happened tonight," Elliot managed, his voice steadier than he felt. "On the balcony. Was that you?"

"That was all you." The stranger leaned forward, his presence somehow filling the mirror. "I just . . . removed certain inhibitions. Showed you what you're capable of when you stop being afraid."

Elliot's throat went dry. "Capable of violence, you mean? I could have broken his wrist. I wanted to break his wrist." The admission sent a chill through him—not just because it was true, but because part of him still wanted it.

"And why shouldn't you?" The stranger's eyes gleamed. "After watching him hurt her for months? The snide comments, the public humiliation, the way he parades other women in front of her?" He shook his head. "The real violence is standing by, doing nothing."

The words hit like a physical blow. How many times had Elliot listened to Maggie cry in the break room? How many times had he offered useless platitudes, knowing she'd go right back to Brad?

"I can't—" Elliot pressed his palms against his eyes until spots danced in the darkness. "This is insane. I'm talking to my reflection about assaulting my friend's boyfriend."

"No." The stranger's voice hardened. "You're talking about protecting someone you love from someone who doesn't deserve her. There's a difference."

When Elliot looked up, the stranger's expression had softened again, understanding and patience radiating from those ancient eyes. "Tell me I'm wrong. Tell me you don't lie awake at night, imagining how different things could be if you just had the courage to show her who you really are."

Elliot's hands trembled against the counter. Every late-night text from Maggie, every tearful confession, every time he'd bitten back what he really wanted to say —it all rose up like bile in his throat. "Even if you're right," he whispered, "even if I . . . I can't just . . ."

"Can't just what?" The stranger tilted his head, studying Elliot like a fascinating specimen. "Can't just stop being her emotional crutch? Can't just show her there's someone who would never treat her the way Brad does?" A knowing smile played at his lips. "You saw her face tonight. That moment of hesitation. She's already starting to see you differently."

The memory of Maggie's tears on the balcony, that charged silence between them, made Elliot's chest ache. But something else nagged at him. "Who are you?" he asked again, stronger this time. "Why are you in my mirror, in my head?"

"Let's just say I have experience with men like you. Men who need a little . . ." the stranger adjusted his collar, and Elliot felt his own hands mirror the movement, ". . . guidance. A little push to become who they're meant to be."

The fluorescent light hummed overhead, casting

strange shadows across their shared reflection. "And what exactly am I meant to be?"

"Someone worthy of her." The words hung in the air like smoke. "Someone strong enough to take what he wants, instead of waiting for scraps of attention."

Elliot's stomach lurched. The stranger's words struck too close to home, peeling back layers he'd carefully maintained. Every friendly lunch, every sympathetic nod, every time he'd swallowed his real feelings—all of it exposed like a raw nerve.

"I don't even know what that means," Elliot said, but his voice wavered. "Being worthy of her. Taking what I want. It sounds . . ."

"Dangerous?" The stranger's smile grew sharp. "Like what you felt tonight, when Brad tried to intimidate you? When you finally stood up for yourself?" He leaned closer to the glass, his presence filling the mirror. "Tell me that didn't feel right."

The memory of Brad's fear, the way power had surged through Elliot's veins—it should have horrified him. Instead, his pulse quickened at the thought. But shame followed close behind, burning in his chest.

"I'm not . . ." Elliot swallowed hard. "I'm not that kind of person."

"No?" The stranger's laugh was soft, knowing. "Then why are you still wearing the jacket?"

The question hit Elliot like a physical blow. He hadn't even thought about taking it off, not once since the balcony. Since that moment with Maggie.

"First things first," the stranger said, straightening his posture. Elliot felt his own spine align, shoulders pulling back. "Those glasses. Take them off."

"They're prescription, if I take them off—"

"Just do it, kid," James said. "Trust me."

Hesitantly, Elliot removed them. He expected everything to get blurry and fuzzy, but instead, everything remained in perfect focus. He looked so different without them, like Clark Kent becoming Superman when the specs came off. "See?" James's approval was almost palpable.

"So now what?"

"Now you ask her out, kid," James said, and the grin on the reflection spread across his face.

"That's—I'm not ready for that. She's not ready for that. Probably never will be."

"There was something there tonight, kid," James said, his voice low and persuasive. "On the balcony. A spark. The way she looked at you, how she hesitated before leaving. You felt it too, didn't you?"

"We're just friends," Elliot protested, but the words sounded hollow. "Work friends. We don't even see each other outside the office."

"Text Maggie, tell her you want to have lunch. This is your chance to show her the real Elliot."

"I don't know . . . tomorrow's Saturday. I'm sure she already has plans and they don't include me."

"Buck up, kid. Have I steered you wrong yet?"

"No."

"Then do what I tell you. Just text her. See what she says." Elliot still hesitated. The phone felt heavy in his hand.

"Sometimes you swing and you miss," James said, his smile knowing. "But sometimes, you hit it out of the park. You'll never know which it'll be until you step up to the plate." He tilted his head. "Show me what you're made of, kid."

Elliot's phone felt like lead in his hand. His thumb hovered over Maggie's contact, heart thundering in his chest. After another moment's hesitation, he typed out the message.

Hey Maggie, you want to have lunch tomorrow?

"There, done. She probably won't even answer."

The response came almost immediately:

Hmm. I've got errands to run tomorrow.

"See, I told you."

"You just giving up already?"

Before Elliot could respond, the phone chimed again.

But maybe we can grab a quick coffee before that. There's a Starbucks at the mall.

"Told you, kid. That's how it's done."

Elliot stared at the phone in disbelief. He reread Maggie's response three times before he looked up at his reflection. "Great. Now I'll probably mess this up and lose her as a friend," Elliot whispered, staring at the screen in disbelief.

"You won't." James's reflection shook his head. "Not with me there. Think of me as your personal advisor. I'll help you find the right words, the right gestures." His smile widened. "After all, I know something about winning a woman's heart."

"Caroline?" The name slipped out before Elliot could stop it, remembered from those whispered suggestions earlier in the night.

A shadow passed across the stranger's face, quick as a blink. Something ancient and pained flickered in those eyes. "My greatest love," he said softly. "But that's a story for another time. Right now, let's focus on Maggie."

Starbucks buzzed with weekend energy—the hiss of espresso machines, the clatter of cups, conversations overlapping into white noise. Normally, Elliot would have hunched his shoulders against the chaos, found the quickest path to the quietest corner. But today, wearing the jacket like armor, he moved through the crowd with an easy confidence that felt both foreign and natural.

He spotted Maggie at a corner table, wrapped around her usual grande chai latte. The morning light from the window caught the amber highlights in her hair, the ones he'd always noticed but never commented on. When she looked up, something flickered across her face, that same hesitation he'd seen on the balcony.

"Hey," she said softly as he approached. "I wasn't sure you'd come after . . . you know. Last night."

"Why wouldn't I?" The smile came easily to his lips, none of his usual self-consciousness present. In the window's reflection, he caught a glimpse of James's approving nod.

She leaned in and gave him a quick, surprising hug. For a second, old habits tried to surface—the awkward tension, the careful distance he always maintained. But the jacket's warmth steadied him, and his arms came up naturally to return the embrace.

When they settled into their seats, Maggie wrapped

her hands around her cup, her eyes meeting his with an intensity that hadn't been there before. "About what happened with Brad . . ."

Easy now, James whispered. *Let her come to you.*

"I couldn't watch him treat you that way anymore." Elliot's voice was gentle but firm, steadied by the jacket's warmth against his shoulders. "You deserve better, Maggie."

She twisted her coffee cup in small circles, her chipped burgundy nail polish catching the morning light. It was a nervous habit he'd never noticed before, or maybe one he'd noticed but filed away with all the other details he wasn't supposed to mention. A flush crept up her neck, coloring her cheeks. "Brad's not always . . . I mean, he can be . . ."

"Cruel?" The word came out softer than he intended, heavy with months of watching, of biting his tongue during lunch breaks when she'd make excuses for Brad's behavior. He shook his head. "The way he grabbed your arm? How he talks over you in meetings? The way he flirts with the interns . . ."

Her eyes snapped to his, something vulnerable and startled in their depths. "You notice all those things?"

"I notice everything about you," Elliot said, surprised by his own honesty.

Maggie's smile faltered slightly. She shifted in her

chair, her shoulders drawing in. But Elliot, caught in the surge of finally speaking these long-held observations, pressed on.

"How your smile never quite reaches your eyes anymore when Brad's—"

"Can we not?" Maggie cut him off softly, her fingers working at the cardboard sleeve of her cup. "Talk about Brad, I mean. That's kind of why I . . ." She gestured vaguely between them. "I just wanted to have coffee with my friend. No drama, no relationship analysis. Just . . . coffee."

The words hit Elliot like a splash of cold water. Of course she'd come here for an escape, and he'd immediately dragged her right back to what she was trying to avoid.

Smooth move, kid, James muttered. *Real smooth.*

"God, I'm sorry," Elliot said, running a hand through his hair. "I didn't mean to—"

"It's okay." Maggie glanced at her watch, then gave him an apologetic smile. "But I should probably get going. I have an appointment with Stephanie at the salon soon."

"Yeah, of course." Elliot watched her gather her purse, feeling the moment slipping away. Their first real interaction outside of work, and he'd managed to make her uncomfortable in record time.

"I wish we had more time," Maggie said, standing. "It was really nice hanging out with you, when we weren't . . ." She let the sentence hang, but her slight smile took the sting out of it.

Now's your chance, James urged. *But tread carefully this time.*

"Yeah, it was nice hanging with you too." The words felt inadequate, but at least they weren't about Brad.

Come on, kid. Don't let it end like this.

Maggie adjusted her purse strap, took a half-step toward the door. "Maybe we can do this again sometime?"

"Yeah, I'd love that," he said, standing as well, feeling the moment dissolving with each backward step she took.

You're swinging and missing, kid. She's giving you an opening.

She gave a small wave, turned toward the door. Something in her posture made Elliot's heart race.

"Uh, Maggie?"

She turned back, eyebrows raised in question. Morning light from the window caught her hair, and for a moment Elliot forgot how to speak.

Ask her, James whispered. *But make it casual. Make it safe.*

Elliot took a breath, his heart hammering against his

ribs. But as he looked at Maggie—her smile genuine but her eyes still carrying that guarded edge—a flicker of doubt crossed his mind. Maybe she was just being polite. Maybe she really did just want to get to her appointment.

Then he remembered her laugh earlier, real and unguarded, before he'd ruined it by bringing up Brad.

"Do you want to hang out at my place tomorrow?" The words came out in a rush. "We could order pizza, watch movies or something?"

Maggie hesitated, her fingers finding the strap of her purse. "Well . . ."

"We won't even talk about he who shall not be named," Elliot added quickly. "Promise."

The tension in her shoulders eased, and this time when she smiled, it reached her eyes. "Yeah," she said, and there was something like relief in her voice. "Yeah, why not? Sounds like fun." She pointed a playful finger at him. "But I pick the movie."

"Of course." His heart felt like it might burst. "Come over at seven?"

"Text me your address." She started backing toward the door, but her smile remained. "I really do have to get to my appointment. Thanks for the coffee and . . ." She paused, something soft in her expression. "Thanks for understanding about . . . you know."

Way to go, kid! James's voice radiated pride. *That's how you do it.*

Elliot stepped out into the morning sunlight, watching Maggie disappear into the weekend crowd. The jacket felt lighter somehow, more like a second skin than a borrowed confidence. For the first time in years, tomorrow felt full of possibility instead of regret.

THE SETTING SUN cast long shadows across Elliot's living room, turning James's reflection in the window into something almost spectral. Elliot adjusted the throw pillows on the sofa for the third time, trying to make them look casually tossed rather than obsessively arranged. The coffee table was a shrine to processed sugar and carbs—chips, pretzels, Red Vines, and enough varieties of popcorn to stock a movie theater. He'd spent forty minutes in Costco, second-guessing every purchase.

"What if she's one of those health-conscious types?" He straightened a bowl of chips that was already perfectly centered. "What if she's gluten-free or—"

Stop fussing, James's voice carried that particular mix of amusement and exasperation Elliot was growing used to. *You've got enough junk food here to feed half the homeless shelters in town. The pillows are fine. The food is fine.*

"I just want—"

Everything to be perfect. I know, kid. But trust me on this, she's into you. A woman doesn't come over for movie night unless she's interested. We just need to play it cool and let nature take its course.

The doorbell's chime sent Elliot's heart rocketing into his throat. Through the peephole, he saw Maggie shifting her weight from foot to foot, wearing fitted jeans and an oversized ISU sweatshirt that somehow made her look both comfortable and devastating. She checked her phone one last time before slipping it into her purse, and Elliot caught the slight tremor in her hands. She was nervous too.

Deep breath, James whispered, his voice a warm current in Elliot's mind. *Remember what we talked about. Follow my lead, and by the end of tonight, everything changes.*

The jacket seemed to adjust itself as Elliot reached for the door handle, leather creaking like a living thing. He felt James's presence settle around his shoulders, a weight both comforting and constraining.

"Hi," Maggie said as the door opened, her smile flickering between genuine warmth and nervous energy. "Hope I'm not late."

"Not at all," Elliot said, stepping back to let her in. The scent of her perfume drifted past. "But I hear it's fashionable to be a little late."

Easy on the corny, James murmured. *She's not here for that.*

"Well, welcome to my place." The words tumbled out before he could stop them, accompanied by a sweeping gesture that felt both too grand and too stiff. Standing in the middle of the living room, he tried to channel some of James's easy confidence. "Let me give you the tour. Living room, obviously. Kitchen through there, and bedroom and bathroom on the right."

"Cozy," Maggie said, her eyes taking in the carefully arranged pillows, the spread of snacks, the subtle signs of his preparation. A smile tugged at her lips. "And clean. I would never have guessed a man lives here."

"I like to keep things organized." The jacket shifted slightly, adjusting to his posture.

"I should have known." She nudged him with her elbow, the casual contact sending electricity through his arm. "I have seen your desk at work."

Don't just stand there like a statue, James urged. *Be a host. Offer her a drink, make her comfortable. Remember what we practiced.*

"Well, come in and make yourself comfortable." His voice came out steadier than he felt.

"Lead the way."

She followed him into the living room, and Elliot found himself hyperaware of her presence behind him, of the soft sound of her footsteps on his carpet.

"Sit wherever you'd like," he said, trying not to watch too intently as she surveyed the seating options.

In the window's reflection, James watched their movements with that ever-present smile, a predator observing its prey. When Maggie settled onto the middle cushion of the sofa, James's whisper carried a note of triumph. *Look at that, kid. No matter where you sit, you're gonna be right next to her.*

"This okay?" Maggie asked, looking up at him. The last rays of sunset caught her hair, creating a halo effect that made Elliot's mouth go dry.

"Uh, yeah, of course." Heat crept up his neck, and he scrambled for his next line. "Hey, do you want a drink? I have soda, water, beer . . ."

"Wine?" Her eyebrows lifted hopefully.

"Yeah, coming right up."

In the kitchen, Elliot retrieved the bottle he'd agonized over at the store—a Cabernet something-or-other that cost exactly $24.99. Not cheap enough to seem thoughtless, not expensive enough to seem trying too hard. At least, that's what he hoped.

As he poured the wine with careful concentration, willing his hands not to shake, he called out, "I ordered the pizza already. Hope you like pepperoni."

"Who doesn't?" Her voice carried easily across the apartment. "Did you order from Pequod's?"

"I did," he said, crossing back to her. Their fingers brushed as he handed her the glass, and the contact lingered a heartbeat longer than necessary. The jacket seemed to warm against his skin.

Maggie took a slow sip of wine, then gestured at his chest with her glass. "Aren't you warm? You're still wearing your coat."

The jacket constricted around his shoulders like a warning. In the window, James's reflection went rigid, that easy smile turning sharp at the edges.

Tell her you're comfortable, James whispered, an edge creeping into his voice. *Change the subject.*

She took another sip, but her eyes never left him. "Take it off, get comfortable, Elliot."

Careful now. James's voice tightened as Elliot's hand moved unconsciously toward the zipper. The leather seemed to writhe against his skin, clinging like a second skin. *You know we can't take it off.*

"I'm comfortable," Elliot said, though the leather

seemed to grow warmer with each passing second. He shifted on the sofa, trying to find a position that didn't make the jacket creak. "So, have you thought about what movie we're watching? It's a big decision."

Maggie tucked her legs under her, the movement bringing her closer. The wine had already brought a slight flush to her cheeks. "I was thinking maybe a horror movie?" Her eyes sparkled with mischief. "Unless you're scared easily?"

She's flirting with you, kid. James's voice was rich with approval. *This is gold.*

"Me? Scared?" Elliot laughed, but sweat was beginning to trickle down his spine, the jacket a furnace against his skin. "Bring it on."

She leaned into his arm as she scrolled through Netflix, her shoulder pressing against the leather. The jacket creaked in response, a sound like teeth grinding together. Elliot felt James's presence intensify, as if drawing closer to the point of contact between them.

Maggie turned to him, concern replacing the playful glint in her eyes. "You're really sweating." She reached toward his sleeve but stopped just short of touching it. "That jacket must be roasting you."

Don't let her make a thing of it, James's voice carried an edge of warning. *Change the subject, quick.*

The doorbell's chime cut through the tension like a

knife. "That'll be the pizza," Elliot said, standing perhaps too quickly. His head swam with heat and relief.

I got this one, kid. Watch and learn.

The change was subtle at first, just a slight heaviness in his limbs as he reached for his wallet. But by the time he opened the door, Elliot felt like a passenger in his own body, like a puppet being guided by invisible strings.

"Uh, that'll be twenty-four fifty," the guy said, holding the pizza box between them like a shield.

Elliot felt his mouth curve into a smile that belonged to someone else. "Here's thirty," he heard himself say, but the voice had James's cadence, James's hint of casual menace. "Keep the change."

See? Smooth as silk. James's voice had changed, taking on a timbre that made Elliot's stomach clench. Something hungry lurked beneath the words, like a predator scenting blood.

Elliot carried the pizza back to the couch, the box suddenly feeling too light in his hands, as if the jacket was doing most of the work. He set it on the coffee table, watching his own movements with a growing sense of disconnection.

"Smells amazing," Maggie said. On the TV screen, the Netflix menu showed "The Conjuring" queued up and ready. "This okay?"

"Perfect." The word came out in his voice, but with

James's inflection. Elliot settled back, electricity shooting through him where Maggie's thigh brushed against his. The jacket creaked as he reached for the pizza box, the sound almost like satisfied laughter.

In the TV's black loading screen, James's reflection appeared, his grin too wide, too sharp. *Look at you, kid. Don't even need my help anymore, do you?* The leather contracted suddenly, pinching the skin at Elliot's neck like possessive fingers.

"Seriously, Elliot." Maggie's touch on his arm was gentle, but her eyes held genuine worry. "You're drenched. At least unzip the jacket?"

She's worried about you. James's voice dripped false sweetness while the jacket's weight pressed down like a lead blanket. *Isn't that sweet?*

Elliot's fingers shook as they found the zipper. The leather twitched and writhed beneath his touch, alive with warning. Like muscle tensing before a strike.

"It's fine," he managed, his smile feeling plastic on his face. "Hey, movie's starting."

Maggie settled against him as the opening credits rolled, her warmth a stark contrast to the cold sweat running down his spine. James's reflection multiplied across every dark or mirrored surface in the room. His encouraging smile stretched impossibly wide, teeth gleaming in the flickering light.

When the first jump scare hit, Maggie grabbed Elliot's arm with a startled laugh. The leather seared his skin where she touched it, like a brand marking territory.

She trusts you, James whispered, his voice thick with satisfaction.

"Seriously, you're really burning up." Maggie's voice was soft with concern as her fingers traced absent patterns on his sleeve. The leather seemed to ripple beneath her touch, like something coiling. "Come on, Elliot, the jacket's got to go."

The zipper pressed into Elliot's chest like a blade as she reached for it. His heart slammed against the leather prison of his ribs, each beat echoing James's growing fury.

Don't you dare, James snarled, and suddenly Elliot's hand wasn't his own anymore. It shot up, catching Maggie's wrist with roughly. His fingers clamped down with a strength that terrified him, moving to a will that wasn't his.

"Ow." Maggie jerked back, confusion bleeding into fear as she stared at his hand. "Elliot?"

"Sorry," he gasped, but his fingers might as well have been carved from stone. "I didn't mean—"

Shut up, kid. James's voice cut through his mind like broken glass. *You're ruining everything.*

The jacket constricted violently, forcing his spine

straight, pulling his shoulders back like a puppet master yanking strings. In the TV's dark screen, James's reflection writhed with rage, his features twisting into something inhuman, teeth lengthening in the darkness.

You think you can do this without me? The voice thundered in Elliot's skull. *You're nothing. NOTHING.*

"Elliot, you're hurting me." Maggie's voice cracked, fear replacing confusion. "What's wrong with you?"

"I'm fine," Elliot tried to say, but what came out was a guttural growl that belonged in a predator's throat. His fingers dug deeper, and he watched in horror as bruises began forming under his grip.

You're ruining everything, James's fury filled every corner of Elliot's mind. *We had a plan. Get close, gain trust, then—*

"Let go." Maggie pulled harder, her free hand braced against his chest. She froze suddenly, her eyes widening as they darted between his face and the jacket. Understanding dawned in her expression, followed by a deeper horror. "Elliot, this isn't you."

The movie flickered across the walls, each shadow becoming a canvas for James's rage. His reflection multiplied and distorted, handsome features twisting into something ancient and cruel.

I chose you because you were weak, James's voice

dripped venom. *Easy to control. But you actually love her, don't you?* A laugh like breaking glass. *Pathetic.*

Elliot strained against his own muscles, but his body had become a foreign country. Sweat poured down his back, the jacket constricting like a python, each breath shorter than the last.

"The jacket," he forced out between clenched teeth, the words burning his throat. "Help—"

SHUT UP!

His free hand snapped up with terrifying speed, fingers finding Maggie's throat. The leather rippled and writhed against his skin, its texture changing, becoming more like scales than hide.

If you won't do what needs to be done, James's growl filled every corner of Elliot's mind, *then I'll do it myself.*

"Please," Elliot gasped as his fingers tightened. "Run."

Maggie's knee drove up hard into his solar plexus. The impact loosened his grip just enough for her to twist free, sending her stumbling backward over the coffee table. The pizza box hit the floor with a wet slap, the sound obscene in the sudden silence.

"What the fuck, Elliot?" Her voice trembled, but her eyes were sharp, calculating. She tracked his movements like prey watching a predator, each step backward bringing her closer to the door.

Get up. James's command pulled at Elliot's spine like hooks in meat. *Don't let her leave.*

His body unfolded from the couch with liquid grace, movements too smooth to be human. The smile that stretched his lips belonged to something that had never known mercy.

"Maggie," he managed to choke out past James's control. "Call the police."

Shut your mouth! The jacket contracted violently, crushing his ribs, stealing his air. His legs carried him forward with inexorable purpose, herding her into the corner. In every reflection, James's teeth grew longer, sharper.

"Stay back." Maggie brandished the lamp like a club, its cord dangling uselessly. Her voice cracked but held steady. "I swear to God, Elliot—"

His arm moved with impossible speed, a cobra strike that sent the lamp flying. It exploded against the wall in a shower of ceramic and shadows. Maggie ducked and rolled, making a desperate break for the door, but his body pivoted with inhuman grace. His hand found her hair, yanked back with practiced cruelty.

"No!" The scream tore from Elliot's throat, raw and desperate as he fought against his own muscles. "Maggie, it's not me, it's the jacket!"

Her elbow shot back, cracking against his ribs with

surprising force. Pain bloomed in his chest, but James's control never wavered.

You really think you can fight me, kid? James's laughter rattled through Elliot's skull. *I've been doing this longer than you've been alive.*

Maggie's next strike found his throat, sharp and vicious. As he staggered, she bolted for the kitchen, knocking chairs aside to block his path.

Amateur. James's contempt was palpable. *Watch how it's done.*

His hand shot out, catching a fistful of her sweatshirt. The yank backward sent them both crashing into the counter, and Elliot felt the exact moment Maggie's fingers found the knife block.

"Don't make me." She brought the chef's knife up between them, its blade catching the TV's flickering light. Her hand shook, but her eyes were steel.

She's got spirit. James's voice dripped with dark pleasure. *This'll be fun.*

"Kill me," Elliot forced out, each word a battle as the jacket constricted like a living straightjacket. "Maggie, you have to—"

His body surged forward. The knife sliced into the leather with a sound like screaming, but the blade didn't penetrate. The jacket had transformed, becoming something ancient and impenetrable. Maggie's eyes went wide as understanding dawned.

My turn.

Elliot's gripped her wrist again, twisting until the knife hit the floor with a damning clatter. His other hand found her throat, lifting her as if she weighed nothing. Her nails raked desperately at the jacket's sleeve, tearing away strips of leather that writhed like living things.

"The zipper," she managed, words barely a whisper. "Elliot . . . help me, pull the zipper . . ."

NO! James's fury exploded through Elliot's mind, but Maggie's fingers had already found the metal tab.

The zipper's teeth separating sounded like a scream of pain. James's howl of rage filled Elliot's skull as Maggie yanked, the leather parting like reluctant flesh.

Cold air hit Elliot's chest like a slap. His fingers spasmed open, dropping Maggie. She hit the floor hard but maintained her grip on the jacket's edges with desperate determination.

STOP HER!

His arms jerked and twitched like a marionette in a storm, trying to grab her, but Maggie held on with terrifying resolve. The leather rippled and twisted, attempting to seal itself shut. She braced one foot against his chest and pulled with renewed strength.

"Fight him!" Her voice was raw but commanding. "Help me get it off!"

The jacket burned like a brand, its leather twisting against his skin. In the reflection, James's human features dissolved into something older, more terrible—a hunger that defied human shape.

You're MINE, James's voice shattered through his consciousness. *You think she'll want you? You're nothing without me. NOTHING!*

Agony ripped through Elliot's body as they forced the jacket wider. He felt James trying to burrow deeper, attempting to fuse their beings into one grotesque whole.

"I'd rather be nothing," Elliot snarled through blood-flecked teeth, "than be you."

The leather peeled away like burning skin, each strip taking surface layers of Elliot with it. His scream harmonized with James's howl of fury as the jacket's grip weakened. James's reflection splintered across every surface, each shard showing his face twisted into something ancient and terrible.

I MADE YOU! James's voice threatened to split Elliot's skull. *YOU WERE PATHETIC BEFORE ME!*

"Keep pulling!" Maggie's voice cut through the chaos like a blade. Her hands moved with surgical precision, methodically stripping away the jacket piece by piece.

Blood ran in rivulets down Elliot's arms where the leather fought to maintain its hold.

His body convulsed violently as James made one final bid for control. The remaining leather constricted with crushing force, but Maggie's grip never faltered.

"I've got you," she said, her voice steel wrapped in velvet. "Fight him, Elliot!"

The final piece tore free with a sound like rending souls. Elliot collapsed, lungs heaving, as James's presence was forcibly extracted from his mind. Between them, the jacket lay on the kitchen floor—not as fabric anymore, but something alive and wounded.

You'll regret this, James's voice echoed from every dark corner. *You could have been—*

"Shut the fuck up," Elliot spat, his fingers finding Maggie's hand with desperate certainty.

The jacket convulsed on the tile, its leather surface bubbling and rippling like skin trying to shed itself. Dark veins appeared across its surface, pulsing with an impossible heartbeat. The sleeves twisted and contorted, reaching blindly for Elliot, for anything to latch onto.

"Jesus Christ," Maggie whispered, pulling Elliot further back against the cabinets.

A seam split along the back, revealing something beneath that wasn't leather at all—a glistening, membranous surface that oozed a thick substance that smelled of copper and rot. The collar folded in on itself repeatedly, like a mouth desperately gasping for air.

The zipper melted and reformed, teeth elongating into actual fangs that gnashed at nothing. Each metal tooth dripped the same dark fluid that was now pooling beneath the writhing garment. In the spreading puddle, Elliot caught reflections—faces of people he didn't recognize, mouths open in silent screams.

"Don't look," he told Maggie, but couldn't pull his own eyes away.

The leather began to crack and peel back, curling like paper in a fire, but slower, more deliberate. Beneath each peeled section was another layer that blackened and disintegrated upon exposure to air. The jacket's movements became more erratic, more desperate, the sleeves batting at the floor as if trying to crawl toward them.

Each piece that broke away dissolved into ash that didn't settle but hung suspended for a moment before dissipating. The jacket's death wasn't clean or quick—it fought for every second, its surface holding James's face for fleeting instants before twisting into something else.

Finally, the remains stopped moving, collapsing inward like a dying star. The leather hardened, cracked, and shattered into fragments that caught the light—not just ash, but crystallized rage, each shard containing a fraction of what James had been. The puddle beneath sizzled and evaporated, leaving dark stains on the tile that formed patterns too deliberate to be random.

In the sudden silence, the only sound was their ragged breathing.

Elliot slumped against the cabinet, his gray t-shirt dark with sweat and blood. Angry red welts marked his skin in patterns like circuit boards where the jacket had tried to make them one. His throat felt scraped raw from screaming.

Maggie knelt beside him, her hands trembling but gentle as she examined the marks. "Jesus," she whispered. Purple bruises were already blooming on her throat where he—where James had grabbed her. "We need to get you to a hospital."

"No hospitals." His voice came out like sandpaper on glass. "How would we even begin to explain this?"

She touched one of the welts and he flinched away. "At least let me clean these."

"Maggie . . ." He caught her hand, his grip gentle now, human. "I'm so sorry. I never meant—"

"Don't." Her voice could have cut steel. "That wasn't you."

They sat in silence, both staring at the pile of ashes on

the floor. In the living room, the movie played on, its manufactured horror now seeming almost laughable.

"I should go," Maggie said, not moving.

"Please don't." The words tumbled out raw and honest. His fingers tightened around hers. "I mean . . . unless you want to. I'd understand if—"

She cut him off before he could ramble. "Shut up, Elliot. I'm not going anywhere."

The ashes stirred in a breeze that shouldn't have existed in the closed kitchen. Elliot kicked at them viciously, scattering them across the floor.

"So," Maggie said after a moment, her voice finding its usual dry humor. "Next movie night, you're wearing a t-shirt."

Elliot laughed, then winced as his ribs protested. "Deal."

"And we're going to talk about this," she added, gesturing at the scattered ashes. "All of it. But first . . ." She stood, helping him carefully to his feet. "First aid kit?"

"Bathroom cabinet."

She supported him as they made their way down the hall, past the devastation of the living room. The shattered lamp, the ruined pizza ground into the carpet, Netflix's "Are you still watching?" message casting judgment from the screen. Everything looked absurdly normal under the apartment's fluorescent lights.

"Sit," she commanded, pointing to the toilet lid. Her hands moved efficiently through his cabinet until she found antiseptic and gauze.

"You don't have to—" he started.

"Shirt off." Her voice brooked no argument. "And if you make one joke about getting me to undress you . . ."

"Wouldn't dream of it." His attempt at a smile twisted into a grimace as he peeled off the sweat-soaked t-shirt. The welts formed an intricate network across his chest and arms, like the jacket had tried to create a roadmap through his veins.

Maggie's touch was clinical but gentle as she cleaned each mark. "You know this is insane, right? Like, actually certifiable insane?"

"Absolutely," he agreed.

He watched her work in the mirror, still struggling to process that she was here, helping him, after everything. "Why did you stay? After . . . after what happened?"

She paused, antiseptic pad hovering over a particularly angry welt. "Because I saw you fighting him. The whole time, even when he had control, you were fighting." Her eyes met his in the mirror, steady and certain. "And because apparently I've been in love with an idiot who needed an evil leather jacket to finally make a move."

"Evil leather jacket was not my first choice for wingman," Elliot said, then hissed as she dabbed antiseptic on a deep welt. "Though he did give decent advice. At first."

"Yeah?" Her hand stilled on his shoulder, warm against his cooling skin. "Like what?"

"Like how I should stop being afraid. How I should tell you . . ." His throat tightened around the words. "Tell you how I felt. How I feel."

She caught his eye in the mirror. "And how do you feel?"

"Sore as hell."

"Elliot."

"Terrified." The word came out raw. "Not of the jacket, or James, or whatever hell that thing was. Terrified you'll walk out that door and never want to see me again."

Maggie set down the antiseptic pad. Her fingers traced one of the welts with butterfly gentleness. "You really think I'd leave? After what we just did?"

"We just fought a possessed jacket."

"No." She turned him to face her, her hands warm on his bare shoulders. "We just fought for each other. Big difference."

A crash from the living room made them both jump.

Maggie grabbed the nearest weapon—a toilet brush—before they crept to the doorway.

The TV lay facedown on the carpet. As they watched, the Netflix screen flickered once, then died with a soft electric sigh.

"Think he's really gone?" Maggie whispered, still gripping the toilet brush like a sword.

"I think . . ." Elliot touched one of the welts on his chest, feeling the raised flesh. "Yeah. I can't feel him anymore. It's just . . . me in here now."

Maggie lowered her makeshift weapon. "Good. Because the whole 'two guys in one body' thing? Major deal breaker."

A laugh bubbled up in Elliot's throat, slightly hysterical. "Noted. No more possessed clothing."

"Or creepy antique shops."

"Definitely no creepy antique shops." He surveyed the destruction in his living room. "I probably won't get my security deposit back."

Maggie snorted, then covered her mouth. Soon they were both laughing, the absurdity of everything finally crashing over them. Elliot's ribs protested, but he couldn't stop.

"Oh God," she wheezed, wiping tears from her eyes. "We should . . . we should probably clean up."

"Yeah." He caught her hand, suddenly needing the contact. "Stay? Help me fix this mess?"

She squeezed his fingers. "Which mess? The apartment or . . ." She gestured between them with her free hand.

"Both?" His voice came out smaller than intended. "I mean, if you want—"

"Elliot." She rose on tiptoe and kissed him softly. "Shut up and get a vacuum."

POLICE BOX
POLICE BOX

"Curious things, telephones. Marvelous inventions that connect us across vast distances, bridging the space between voices with nothing but wire and signal. But what of the distances that can't be measured in miles? What of the ultimate separation?

"This particular piece caught Chelle's eye—a whimsical fusion of science fiction and mythology. A TARDIS ensnared by a Kraken's tentacles, reimagined as a telephone. She found it charming, a perfect addi-

tion to her collection of oddities. Twenty pounds seemed a small price for such a conversation piece.

"Chelle understands the power of voices better than most. Night after night at the Samaritans center, she listens to strangers in their darkest moments, offering comfort through nothing but her words across the line. Ten years she's been doing this work—ever since her sister Katherine's passing.

"How interesting that she should be drawn to this particular item. As if something within it called specifically to her. As if it knew exactly what connection she most desperately wished to restore.

"But I wonder . . . if the dead could speak to us again, would their words bring comfort or terror?

"Some calls, once connected, cannot easily be disconnected. Some voices, once invited in, refuse to remain confined to mere telephone lines.

"Perhaps you'll take it home and discover for yourself. The price remains twenty pounds. Cash only, please."

Chelle pushed through the double doors of the Samaritans center, the predawn air hitting her face like a splash of cold water. Her shoulders ached from sitting too long, but her mind was still with her last caller, a widow who couldn't sleep, who just needed someone to listen while she talked about her husband's garden and how the weeds were taking over now. Sometimes just being there, really being there, was enough.

The night shift had been steady: a university student panicking over exams, a man struggling with his divorce, a teenager who felt alone. No crisis calls tonight, thank God, but every call mattered. Every voice deserved to be heard.

The streets of Lancaster were empty except for the occasional delivery van, though the bakery two blocks down was already warming up for the morning rush, scenting the air with fresh bread. Chelle pulled her cardigan tighter, wishing she'd brought a proper coat.

She was already mentally planning the rest of her day —sleep until whenever, then maybe the pub for a late lunch. The Crown did proper chips on Thursdays, none of that frozen rubbish. The thought of food made her stomach growl, reminding her she'd skipped dinner during her shift.

The shops that lined the street were all shuttered

this early, metal grilles pulled down tight. Except there was light spilling from one of the storefronts. The Charity Shop—Grand Opening the sign in the window announced in fresh gold lettering. Chelle slowed. She could have sworn this was empty yesterday, with a "To Let" sign gathering dust in the window.

As she passed, something in the display caught her eye—a TARDIS being attacked by what looked like a Kraken, its tentacles wrapping around the blue police box. She moved closer to the glass for a better look. A smile spread across her lips and she found herself entering the shop, as if guided on autopilot.

Inside, the shop was warmer than she expected, with that peculiar mix of dust and old fabric that all charity shops seemed to share. A bell tinkled above her head as the door closed behind her. The place was cramped but tidy, with racks of clothes, shelves of books, the usual bric-a-brac arranged on tables.

The TARDIS-Kraken piece sat in the window display where she'd spotted it. Up close, it was even better—the detail work on both the police box and the creature was impressive. What she'd thought was just a novelty sculpture turned out to be a telephone, with the handset cleverly worked into the design. A small price tag dangled from it: £20.

"Ah, a fan of Doctor Who or sea monsters? Or both, perhaps?"

Chelle startled. Behind her stood a tall, thin person wearing what looked like an outdated suit. Their smile was knowing, almost amused.

"Both, actually."

"The creator had quite the imagination," they continued, moving behind the counter. "Combining two seemingly unrelated things into something rather unique. Though sometimes the most unexpected combinations yield the most interesting results, wouldn't you agree?"

"It's perfect," Chelle said, already reaching for her wallet. No way was she leaving without it.

The shopkeeper's smile widened slightly as they wrapped the phone in brown paper. "I do hope it brings you some interesting conversations."

"Does it even work?" Chelle asked, wrinkling her nose and taking the wrapped package.

"Oh, it works perfectly well for the right kind of calls," the shopkeeper's eyes glinted. "The ones that really need to get through."

Something in their tone made Chelle hesitate, her fingers tightening on the paper. "Right. Well, it's just for decoration anyway."

Outside, the morning was getting brighter, though the chill hadn't lifted. Chelle tucked the package under her arm and started the ten-minute walk to her flat. Her mind drifted between thoughts of sleep and wondering where she'd display her new find. Maybe on the bookshelf in the living room, or perhaps her bedroom desk where she could see it while reading.

The weight of the night shift was settling into her bones now. Eight hours of listening, really listening, took more out of you than people realized. But it was worth it. Even on the hardest nights, knowing she might have made someone feel less alone made it worth it.

She climbed the stairs to her second-floor flat, fishing her keys from her bag. Inside, she dropped her bag by the door and kicked off her shoes. The package went on her kitchen counter while she made a quick cup of tea—she never could sleep right after a shift without unwinding first.

The kettle boiled as she unwrapped the phone, grinning at how ridiculous and perfect it was. The TARDIS portion lifted up as the handset, painted that familiar police-box blue with tiny white-framed windows and the "Pull to Open" sign clearly visible. The Kraken base was a masterpiece of detail, tentacles curved and twisted to create the cradle, each sucker precisely carved before ending in the number buttons. The creature's body formed the base, its eye seeming to glare balefully at the TARDIS it had captured.

It was exactly the kind of weird, wonderful thing she loved. The designer must have been either brilliantly creative or completely mad, probably both. She could already imagine her friends' reactions when they saw it. Jules would roll her eyes but secretly love it, and Matt would probably try to convince her to sell it to him—he was always trying to go one better than her.

The kettle clicked off and Chelle made her tea, adding just a splash of milk. She settled onto her worn leather couch, feet tucked under her, and sipped while studying her new purchase. The craftsmanship really was remarkable—even the TARDIS's wood grain had been carefully etched into the handset. She lifted it experimentally. Despite its ornate design, it balanced perfectly in her hand.

Her flat was quiet except for the occasional rumble of early morning traffic and the soft tick of the kitchen clock. Usually after a night shift, she'd watch something mindless on Netflix to wind down, but her eyes were already heavy. The tea was working its magic, melting away the tension from sitting too long in those uncomfortable chairs.

She should probably shower, but bed was calling. Setting her empty mug in the sink, she carried the phone into her bedroom. Her collection of cryptozoology books lined the shelves above her desk—everything from

Bigfoot to the Loch Ness Monster. The TARDIS/Kraken phone would fit right in. She cleared a space between "Sea Monsters: A Visual History" and her dog-eared copy of "British Folklore and Cryptids," positioning it so the TARDIS faced out toward her bed.

Perfect. She could almost hear River Song saying, "Hello, Sweetie" to the captured police box.

Chelle changed into an oversized t-shirt and flannel shorts, tossed her clothes into the hamper, and went through her abbreviated routine on autopilot. Teeth brushed, face washed, done. Her hair was a mess but whatever.

She crossed the room and pulled the heavy blackout curtains shut against the brightening morning. The only light came from her small bedside lamp, casting a warm glow over her familiar space. Movie posters, mostly horror and sci-fi, covered one wall. A large cork board dominated another, pinned with photos of friends and family, ticket stubs, and other memorabilia.

Chelle clicked off the lamp and crawled under her duvet, the cotton cool against her skin. From where her head rested on her pillow, she could just make out the silhouette of the TARDIS telephone on her desk. Her last coherent thought before drifting off was that she should text Jules a picture of it. Her best friend would definitely appreciate the weirdness.

THE LATE AFTERNOON sun slanted across the street as Chelle headed toward The Crown, feeling more human after coffee and a quick change. She'd pulled on her favorite jeans and a vintage Clash t-shirt, hair still messy but artfully so. The spring air had warmed considerably since dawn, though a brisk breeze still whipped around corners, carrying the scent of fresh-cut grass from the park.

The Crown was doing steady business—the usual mix of students and locals filled the worn wooden tables. Behind the bar, Pete was pulling pints with practiced efficiency. The ceiling fans spun lazily, stirring the perpetual haze of conversation and classic rock playing just loud enough to fill the gaps.

"Alright, love?" Pete called as she claimed her usual spot at the bar. "Night shift again?"

"Yeah," she said, sliding onto the stool. "Fish and chips still on?"

"Course. Mushy peas?"

"Is there any other way?"

While Pete put her order in, Chelle surveyed the pub. The pool table in the back was free, unusual for this time of day. A few regulars nodded hello. The Crown was that kind of place, where everyone knew everyone else but respected the unwritten rule of leaving you alone unless invited to chat.

Her food arrived steaming, the batter golden and crisp, chips thick-cut and properly done. Chelle doused everything in malt vinegar, watching it soak into the paper. The first bite was perfect—flaky white cod under that crunchy exterior. This was why she kept coming back despite the slightly sticky tables and the ancient jukebox that only played half its selections.

Jules showed up just as she was finishing, dropping onto the next stool with her usual dramatic flair. Her friend's dark curls were tied up in a messy bun, paint stains visible on her hands.

"You look suspiciously awake for someone who just did nights," Jules said, stealing a chip. "Oh, and you have to see what this prat tried to submit for the spring exhibition."

Chelle let Jules ramble about incompetent artists while she finished her food, then nodded toward the pool table. "Fancy a game? Loser buys the next round."

"Only if you promise not to do that thing where you pretend to be rubbish for the first three shots."

They claimed the table, racking the balls while Jules continued her story about the would-be artist and his installation made entirely of broken electronics. Chelle chalked her cue, grinning. She hadn't mentioned her strange purchase yet—better to save that for when Jules was properly distracted by the game.

Jules broke, scattering the balls without sinking any. "Bugger." She stepped back, gesturing dramatically at the table. "Your turn, shark."

Chelle lined up her shot, easily dropping the 3-ball in the corner pocket. "Not my fault you never learned proper technique." She circled the table, looking for her next shot. "You'll never guess what I found at the new charity shop that opened up on Preston Street."

"Since when is there a charity shop on Preston?" Jules chalked her cue, frowning. "Thought that was all estate agents and coffee places now."

"Grand opening, apparently." Chelle sank another ball, then missed her third shot. "Got this mad phone thing. TARDIS being attacked by a Kraken. Proper detailed too, all Victorian-looking."

Jules perked up, nearly missing her shot. "Pictures or it didn't happen."

"On my phone." Chelle pulled it out, then paused. "Hang on, let me get the next round first. Same?"

"Half of Horizon, thanks. And you better not be winding me up about this phone thing. You know how I feel about weird art."

Chelle returned with their drinks—a half of Horizon pale ale for Jules and a pint of Blackbird stout for herself. Jules was studying the table with exaggerated concentration, probably planning some ridiculous bank shot that would never work.

"Right, look at this," Chelle said, pulling up her photos. Except . . . she hadn't taken any yet. "Oh, forgot I was too knackered this morning. I'll send you one later."

"Convenient," Jules said, attempting and spectacularly failing her bank shot. "Sure this mystery shop even exists?"

"Course it does. Next to that vegan café, you know, the one with the awful puns on their sandwich board." Chelle lined up an easy shot on the 7-ball. "Speaking of which, how's the new exhibition coming along? Still fighting with that lighting setup?"

Jules launched into a detailed complaint about gallery spotlights and color temperature, gesturing with her cue until Chelle had to duck to avoid being accidentally smacked. The game continued, their conversation drifting between Jules' work, Chelle's latest true-crime podcast obsession, and whether they should try that new Korean place next week.

"And that," Chelle said, sinking the black with a satisfying thunk, "is why you shouldn't bet against someone who misspent their university years in this very pub."

Jules groaned, setting her cue in the rack. "One day you'll teach me how you do that thing with the spin."

"Never. A magician never reveals their secrets." Chelle checked her phone. It was nearly eight. "Think I'm going to head home, have a proper soak in the tub. Been looking forward to finishing that new Gothic horror novel."

"The one with the haunted lighthouse?"

"That's the one. You can borrow it when I'm done." Chelle pulled on her cardigan. "Thanks for the last round, by the way."

They hugged goodbye outside The Crown, the street lamps just flickering to life. The evening air had that peculiar spring quality, warm but with an edge of lingering winter. Chelle turned toward home, her thoughts already drifting to hot bath water and her waiting book.

She was passing the vegan café when she noticed something odd. Where the charity shop had been earlier today was nothing but papered-over windows, a "To Let" sign hanging crookedly in the corner.

Chelle stopped, frowning. The sign was weathered, the papers yellow with age. Not the work of a single day. She touched the window and dust came away on her fingertips.

A couple hurried past, forcing her to step aside. The street was filling with the usual Thursday evening crowd, people spilling out of bars and restaurants. She could still picture the bright "Grand Opening" banner, the gleaming display window, that odd shopkeeper with their knowing smile.

Her hand went to her phone, thinking to text Jules, then stopped. The spring breeze had turned cold now, finding its way under her cardigan. Chelle pulled it tighter and started walking again. What she needed was that hot bath, maybe with a glass of wine.

HER FLAT WELCOMED her with familiar shadows and silence. Chelle kicked off her shoes, hung her cardigan, and headed straight for the bathroom. The old pipes groaned as she turned on the taps, steam quickly fogging the mirror. She added a generous splash of lavender bath oil—a Christmas gift from Jules that had turned into a proper addiction.

While the tub filled, she poured a glass of the Malbec she'd been saving. The Gothic novel waited on the bath caddy, its bookmark promising the start of a new chapter. It was starting to get to the good bits where the protagonist had just discovered the lighthouse keeper's journal.

Chelle twisted her hair into a messy top knot and sank into the hot water with a contented sigh. The wine was rich and warming, the book properly atmospheric. Through the bathroom window, she could hear the distant sounds of the city settling into evening—traffic, scattered voices, the last birds finding their roosts.

The mystery of the vanishing shop tickled at the edge of her thoughts, but the heat and wine were doing their work, melting away the lingering confusion. She turned another page, losing herself in tales of stormy seas and mysterious lights.

She was three more chapters in when her wine glass reached empty. The water had cooled just enough to need a top-up of hot. Chelle marked her place with the book-mark—the lighthouse keeper had just discovered strange markings carved into the basement walls—and sat up to adjust the taps.

The flat was utterly quiet now, that peculiar stillness when the world outside seems to hold its breath. The lavender scent had mellowed, mixing with the familiar bathroom smells of her shampoo and the vanilla candle burning on the windowsill.

She poured more wine from the bottle and had just settled back, reaching for her book, when she heard it. A sound so familiar from countless episodes of Doctor Who that her brain almost dismissed it as imagination—the distinctive wheezing, groaning materialization of the TARDIS.

And then, immediately after, something else—a deep, resonant cry that made her spine stiffen. She held her breath, listening intently. Then it came again, that same sequence. The TARDIS's distinctive wheezing groan, followed by the bone-deep cry of something ancient and vast.

Water sloshed over the edge of the tub as she stood up, grabbing her towel. Chelle stood dripping on the bathroom tiles, water pooling at her feet. The sounds came yet again, the TARDIS, then what now dawned on her was the cry of the Kraken. She actually laughed, a short surprised sound that echoed off the bathroom walls.

Of course the bloody thing made noises. Because why wouldn't a telephone that looked like a TARDIS being attacked by a Kraken also sound like one?

"Alright, you weird thing," she muttered, adjusting her towel more securely. "Let's see if you've got an off switch."

She padded down the hallway, leaving wet footprints on the floor. The sounds were getting louder. Whoever designed it had done a proper job with the audio. It was actually quite cool, though maybe not at—she glanced at her watch—quarter past nine on a Thursday evening.

The phone sat on her desk exactly where she'd left it.

The TARDIS sound came again, followed by the Kraken. She reached for it, looking for any kind of switch or button that might silence it.

The TARDIS materialization echoed again, followed by that deep Kraken cry. Chelle turned the phone over carefully on her desk, water still dripping from her and onto the wooden surface. The craftsmanship really was remarkable—every tiny detail perfect, from the tentacles wrapping around the TARDIS to hand-carved white letters.

"Come on then," she murmured, running her fingers along the base. "Where's your battery compartment or switch or whatever?"

She turned it over in her hands, still searching for a way to shut the thing up.

"Proper weird, you are," she said to the phone. "Though I suppose that's why I bought you."

Finally, she reached for the TARDIS-shaped handset, and lifted it away from where it sat nestled in the Kraken's carved tentacles, trying to make it stop. As she examined the miniature TARDIS for an off switch, a soft voice came through clearly:

"Hello? Is . . . is anyone there?"

Chelle nearly dropped the handset.

"HELLO?" Chelle said into the TARDIS handset, feeling ridiculous standing there dripping bathwater onto her desk. The carved phone wasn't even connected to anything—just a decorative piece that somehow made sounds.

"Hello?" the voice came through with startling clarity, as though carried on a perfect telephone line. "Can anyone hear me?"

Chelle's breath caught. The voice had an odd quality to it, like it was coming from very far away, yet somehow immediate and present—as if someone were whispering directly into her ear. A woman's voice.

"Yes, I can hear you," Chelle said carefully, unconsciously shifting into the measured tones she used on her Samaritans calls. Her fingers tightened around the TARDIS as water trickled down her spine, raising goosebumps in its wake.

For a moment there was no response, only breathing from the other end—soft, hesitant breaths that seemed to fill the silence of her bedroom.

"Who are you trying to reach, love?"

"I . . . I'm not sure . . ." The voice trailed off, uncertain, like someone waking from a deep sleep.

"You're not sure who you're ringing?" Chelle asked, even as her heart began to race. Something about that voice tugged at her memory.

"I was ringing Chelle?" The voice cracked, a familiar tremor that made her stomach twist.

Chelle's hand tightened on the TARDIS. Water dripped steadily onto the polished wood of her desk as she stood frozen, trying to process what was happening. Each drop echoed in the quiet room like a tiny heartbeat.

"I'm Chelle," she said, then swallowed hard. "Who is this?"

A shaky breath came through the TARDIS handset, now so clear it could have been someone standing right next to her. "You probably won't believe me. I'm not sure I believe this is happening myself."

"You need to tell me who you are," Chelle said, trying to keep her voice steady. Her free hand gripped the edge of the desk, anchoring herself in the solid reality of wet wood under her fingers.

"It's . . ." Another shaky breath through the handset. "It's Katherine. Your sister."

The words hung in the air between them, heavy as storm clouds. Chelle's legs felt suddenly weak, and she sank into her desk chair.

"That's not possible," she whispered, but even as she said it, she knew she recognized the voice, now that it was clear. Knew it as surely as she knew her own reflection. "Katherine's dead. She died ten years ago."

"I know," the voice cracked again. "I know I did. But Chelle . . . Chelleybean . . . I need to talk to you. Please don't ring off."

That nickname. Nobody had called her that except Katherine. It wrapped around her heart like a familiar embrace.

"How do I know this is real?" she managed finally. "How do I know I haven't just . . . just finally gone round the bend? Or that this is some sort of sick joke?"

"Do you remember that summer in Cornwall?" Katherine's voice came soft through the handset, gentle as sea spray. "When I was seventeen and you were twelve? You begged me to let you climb the rocks at Tintagel and dad got so cross."

Chelle's throat tightened. She remembered the salt spray, Katherine's hands steady on her back as she reached for the next handhold. Their father's face when he spotted them from the beach below—terror and fury warring in his expression.

"You told him it was your idea," Chelle said, barely audible. "You always did that. Took the blame when I got in trouble."

"Had to look after my little sister, didn't I?" A familiar laugh, warm and bright as summer sunshine. "God, Chelle, I've missed you so much."

Chelle sat there, clutching the ridiculous handset, trying to wrap her mind around what was happening. This was impossible, yet it was. She was sitting here, wrapped in a bath towel and talking to her dead sister on a novelty phone that wasn't even plugged in.

"Where . . ." she swallowed hard, forcing the words out. "Where are you, Kath?"

"I'm not certain," Katherine's voice wavered, like a radio signal threatening to fade. "It's dark here. Bloody freezing. I've been trying to reach you for ages . . ."

"Trying to reach me?" Chelle's voice cracked. She wiped water from her face, tears now, not just bathwater. "Why now? Why after all this time?"

There was a long pause. The TARDIS grew warm in her hand, the sharp corners of it scoring lines into her palm—the only thing that felt real in this impossible moment.

"Because I need to tell you something," Katherine said finally. "About the night I died."

"Why didn't you ring me then, Kath?" Her voice sounded small, young—like the child she'd been when they shared secrets in the dark. "All these years, I've wondered . . . why didn't you reach out? Just a quick call, anything."

"Would you believe me if I told you I tried?" Katherine's voice had that soft, broken quality Chelle remembered from after the accident—like shattered glass barely held together. "Sat there for ages, just . . . staring at your number on my mobile."

"But you never rang." Bitterness crept into Chelle's voice, surprising her. She pressed the police box phone harder against her ear, as if she could somehow reach through time and space to grab her sister, shake some sense into her. The sharp edges of the TARDIS dug into her palm like accusation. "I would've helped you. Christ, Kath, I would've—"

"That's exactly why I couldn't, wasn't it?" Katherine's words cut like glass. "Because you'd have tried your hardest to save me. Known exactly what to say, exactly how to talk me round. And I just . . . I couldn't let you."

Chelle's fingers ached from gripping the square handset, but she couldn't make herself let go. The pain felt like penance. "I don't understand. Why wouldn't you want me to help?"

Katherine's laugh was hollow, empty—a sound Chelle had never heard from her sister in life. It echoed through the handset like wind through an abandoned house. "Because I didn't bloody deserve it, did I?"

"What are you on about—"

"Stace and Martin." The names fell like stones into the silence, heavy with the weight of a decade's guilt.

Chelle's stomach lurched. The crash. God, she could still remember every detail of that night—the chaotic hospital corridors, Katherine pale in the bed, bruised and scratched but alive. And the news about her mates . . .

"I was the one behind the wheel," Katherine whispered, her voice thick with remembered horror. "Pissed out of my mind, wasn't I? But somehow, I was the only one who walked away."

"It wasn't your fault, Kath," she said, the words feeling worn and inadequate after ten years of silence.

"Wasn't it?" That hollow laugh again. "I chose to drive, didn't I? Then had to choose to live, every bloody day after. Watching their parents at the funeral. Walking past their empty desks at sixth form. D'you know what that's like, Chelleybean? Living when you should've died?"

The childhood nickname twisted in Chelle's gut like a knife. She slumped back in her chair, free hand pressed against her eyes, phone burning against her ear like a brand. "So that night . . ."

"I wanted to hear your voice." Katherine's words were barely a whisper now. "One last time. But I knew what would happen. You'd have cottoned on straight away—you always could read me better than anyone. You'd have said all the right things, made all the right arguments. Made me want to stay."

"And that would've been wrong?" Chelle's voice cracked.

"Would've been selfish, wouldn't it? Carrying on just because I was too weak to face what I'd done. Letting you talk me into a future I had no right to." Katherine's voice seemed to fade, growing distant as though being pulled away by an invisible tide. "I was knackered, Chelle. Just . . . properly done in from surviving when they couldn't."

"Talk to me now, yeah?" Chelle said softly, deliberately. "Properly talk to me. Like you couldn't that night."

"I don't know where to begin," Katherine's voice wavered through the handset. "It's been ages, and there's so much I ought to have said."

"Start anywhere," Chelle said gently, using the calm, steady tone she'd learned through countless Samaritans calls—though her hand trembled against the absurd TARDIS telephone.

"I've watched you, you know," Katherine said softly, something different creeping into her voice now—something careful, measured. "All these years. Watched you tear yourself to bits thinking you should have noticed, should have known. Watched you spend every Saturday night at that crisis center, trying to save everyone else because you reckon you failed to save me."

"Didn't I? Fail you?"

"No," Katherine's voice was firm now, that familiar big-sister tone, but somehow wrong—like an echo of itself. "You were fourteen, Chelle. I was the one who hid it. I got so bloody good at pretending everything was fine that sometimes I almost believed it myself. Until I couldn't anymore."

Tears mixed with the bathwater on Chelle's cheeks, both warm and cold against her skin. "Why didn't you let me help you, Kath? I would have tried so hard."

"That's exactly why," Katherine whispered, her voice soft as a confession in church. "Because you would have tried. You'd have taken on that burden too, on top of everything else. And I couldn't do that to you. Not then."

"But you can do it now?" Chelle asked, her voice barely steady, an edge of something sharper than grief creeping in. "After ten years of . . . of watching me? Why now, Kath?"

The TARDIS grew uncomfortable against her palm, its edges seeming to press deeper with each passing moment. Through it, Katherine's breath came in that familiar pattern, the one that meant she was holding back tears—a rhythm Chelle had memorized in childhood and never forgotten.

"Because I need you to let it go," Katherine said finally, each word measured and precise, like stepping stones across dark water. "You've built your whole life around that night. Around trying to save everyone else because you reckon you failed to save me. But my death wasn't your responsibility, Chelleybean. It was never your burden to carry."

"How can I let it go?" Chelle whispered, the words barely disturbing the air. "Every time I take a call at Samaritans, every time I help someone . . . I reckon maybe this makes up for not helping you. Maybe this time—"

"That's not why you ought to be doing it," Katherine cut in, her voice taking on an odd urgency that made the phone warm further in Chelle's grip. "Help people because you want to, because you're brilliant at it. Not because you're trying to make up for something that wasn't your fault to begin with."

The words hung in the air between them, heavy with ten years of unspoken grief. Chelle's fingers trembled against the uncomfortable shape of the handset, its blue paint now feeling almost feverish against her skin. The silence stretched between them like a thread pulled too tight, ready to snap.

"Not my fault," Chelle repeated softly, her fingers tracing the blue wooden panels of the police box phone in her hands. The grain of the wood felt different now—more alive somehow, thrumming beneath her touch. "Is that why you're ringing? To absolve me?"

"I'm ringing because . . ." Katherine's voice faded slightly, like a dodgy connection, before strengthening again with an odd clarity that made Chelle's skin prickle. "Because this phone, whatever it is . . . it's given me a chance to reach back. To properly talk when we're both ready to hear each other."

"I hear you," Chelle said, wiping tears from her cheeks with her free hand, the moisture cool against her heated skin. "But Kath, all these years of watching . . . you've seen what I do now. How many people I've helped. Are you saying that meant nothing?"

"It meant everything," Katherine's voice softened, though something else threaded through it now—an undertone of urgency that made the phone pulse warmer in Chelle's grip. "But you did it for the wrong reasons, didn't you? You turned your grief into a mission, your guilt into a purpose. I'm dead chuffed with who you've become, but I need you to become it for yourself, not for me."

Chelle clutched the impossible phone tighter, this strange little police box that could somehow reach across the years, letting her sister say what she never could in life. The edges pressed into her palm with an almost hungry insistence now, as if trying to merge with her flesh.

"I need you to listen to me now, Chelle," Katherine's voice took on a new urgency, sharp and immediate. "Properly listen."

"I'm listening, Kath. I'm right here." The words fell into the growing tension like stones into still water, rippling through the quiet of her bedroom.

"You were fourteen when I died. Fourteen." Katherine's voice cracked. "And you've spent ten years trying to save me, over and over, in every stranger you talk to. But you can't save someone who's already gone, can you? You can only . . . ."

"Only what?"

"Save yourself," Katherine whispered. "Let me go, Chelleybean. Let the guilt go. I need you to—"

Static crackled through the line, Katherine's voice suddenly fading in and out like a dodgy radio signal. Chelle pressed the TARDIS harder against her ear as she hunched forward.

"Kath? You're breaking up, I can't—"

"Running out of time," Katherine's voice came back stronger, more desperate, with an edge that made Chelle's skin prickle. "Please, Chelle, you've got to understand. I never blamed you. Never. But something's coming, and you need to—"

"Need to what?" Chelle asked, adjusting her grip on the phone, its surface now uncomfortably warm against her palm. "What are you trying to tell me, Kath?"

"That watching you punish yourself . . . it's bloody awful," Katherine's voice softened, though something lurked beneath the gentleness. "Every Saturday night at that crisis center, every call you take . . . I see you trying to save me through them."

"I help people," Chelle whispered. "That's got to count for something."

"Course it does. But Chelle . . ." Katherine paused, that familiar hesitation before saying something difficult, though now it felt rehearsed, calculated. "You're brilliant at it. Absolutely natural at helping people through their darkest moments. But you do it like it's penance, don't you? Like each life you save might somehow balance out losing me."

The truth of it hit Chelle hard. Every late-night call, every desperate voice she'd talked back from the edge, she'd been trying to rewrite that one night, over and over.

"I don't know how to do it differently," she admitted, her voice small.

"You start by forgiving yourself," Katherine said gently, though her voice had taken on an odd cadence. "By understanding that what I did wasn't your fault. That fourteen-year-old you didn't fail anyone, love."

"But I still help people. That bit doesn't change, yeah?"

"Course not. But you help them because you can, because you're brilliant at it. Not because you're trying to save me through them."

The blue wooden phone hummed slightly, a low vibration that seemed to travel up Chelle's arm, and Katherine's voice took on an odd echo. "Time's strange, innit? Sitting here, talking to you like this . . ."

"Through an impossible phone," Chelle said, running her fingers over the police box panels, their edges now sharp enough to leave impressions in her skin. "Kath . . . how are you even here? Really here, I mean?"

A long pause followed, heavy with something unspoken. When Katherine spoke again, her voice had changed subtly, like an instrument slightly out of tune. "Does it matter? Isn't it enough that we're talking?"

"It matters," Chelle said slowly, her crisis training picking up something off in her sister's tone. Something that hadn't been there moments before. "Because I need to know if this is real, or if I'm . . ."

"If you're what, Chelleybean?" The childhood nickname sounded wrong somehow. Hollow. Like something wearing her sister's voice.

"If I'm just hearing what I want to hear," Chelle

finished, gripping the TARDIS tighter, its heat now almost unbearable.

"You always were too clever," Katherine said softly. "Too good at sussing people out. Even me. Especially me."

"Kath—"

"But that's why I couldn't ring that night, remember? Because you'd have known. You'd have heard it in my voice."

"You've already told me that bit," Chelle said carefully.

"Did I?" Katherine's voice wavered slightly, like a record playing at the wrong speed. "Time's gone all wonky like this. Everything bleeding together. But you understand now, don't you? Why I'm here?"

Chelle stood slowly, water cascading off her body, each droplet seeming to echo in the growing tension. "I thought I did. But now I'm not sure."

"You don't trust me?" Katherine's voice held an unfamiliar note, something ancient and hungry beneath the hurt. "After all this time, after finally reaching you . . ."

"Course I trust you," Chelle said, every movement suddenly feeling like it needed to be careful, measured. "But something feels . . ."

"Wrong?" Katherine finished. "Because I'm dead? Because this is bloody impossible?"

"No," Chelle adjusted the towel around herself as she shifted in the chair. "Because you keep circling back. Like you're trying to make sure I've cottoned on to something."

The silence stretched between them, broken only by that too-regular breathing on the other end of the line— mechanical, like a metronome counting down.

"I just want to help you," Katherine said finally. Her voice sounded further away somehow, despite the phone's increasing warmth against Chelle's ear. "Like you help all those people at Samaritans. Like you wanted to help me."

A sharp crack split the silence—like splintering wood —somewhere in her flat. Chelle jolted upright, her heart hammering against her ribs.

"Chelle, still there, yeah?"

"Still here. Thought I heard something." Chelle's whisper barely disturbed the air around her.

"Heard what?" Katherine's voice had changed again, lighter, almost playful. The shift felt wrong, like watching a familiar face distort in a funhouse mirror. "Are you expecting company, Chelleybean?"

That childhood nickname twisted wrong in Chelle's gut. The way Katherine said it now felt like someone trying on borrowed clothes, mimicking a familiar gesture without quite getting it right—a stranger wearing her sister's voice like an ill-fitting mask.

"No," Chelle said carefully. The phone cut into her palm, its heat now almost blistering, but she couldn't make herself let go. "No visitors. Not at this hour."

"Then perhaps you ought to check," Katherine suggested. "Make sure everything's alright. You can take me with you."

Another impact thundered through the flat, closer this time—a heavy, deliberate thud that seemed to echo in Chelle's bones. Her fingers tightened around the impossible TARDIS telephone.

"Kath," she said, forcing her voice steady like she did during difficult calls. "Earlier, you said you were watching me. For ten years. What did you mean by that?"

The breathing on the other end of the line stopped entirely for three heartbeats before resuming its mechanical rhythm, like a ventilator keeping something alive that should have died long ago.

"I told you," Katherine said, but the voice wasn't quite her sister's anymore. "I needed to make sure you understood."

"Understood what?"

A third crash reverberated through the walls—methodical, purposeful—the sound of something moving through her flat, testing each door with devastating patience.

"That it wasn't your fault," Katherine's voice had taken on an odd cadence, like a record playing at the wrong speed. "That you need to let go."

"You keep saying that but you're not really telling me anything new, are you? You're just . . . going round in circles."

The mechanical breathing hitched, like a machine struggling against rust. "I'm trying to help you."

"No," Chelle whispered, her crisis training finally clicking into place. The way you learned to spot circular conversations, manipulative patterns. "You're trying to keep me talking. Keeping me on the line until—"

A knock at her bedroom door cut through the silence. Three taps, like someone using a single knuckle against wood—sharp, deliberate.

"Aren't you going to answer it?" Katherine asked.

Chelle stared at her bedroom door, the blue police box phone burning in her grip like a brand.

The knock came again. Three precise taps. Patient. Waiting. The sound seemed to ripple through the air, each impact leaving aftershocks in its wake.

"Kath," Chelle's voice cracked as she stared at her bedroom door. "What's outside the door?"

The breathing on the phone stuttered, like a skipping CD caught in an endless loop.

The knocking came again, but different this time, not three precise taps, but a child's rhythm. Shave and a haircut. The playful pattern felt obscene in the heavy silence.

Chelle's blood ran cold. It was the secret knock they'd used as kids.

"Please, Chelleybean." The voice on the phone said. "Let me in."

From the other side of her bedroom door, in that same beloved voice: "Let me in, Chelleybean."

Chelle stood, her legs trembling slightly as she set down the TARDIS and pulled the towel more securely around her body. She stared at the door, her sister's voice echoing from both the phone and the hallway beyond. Everything she knew screamed at her to leave the door shut.

Yet her legs carried her forward before she could stop herself. The rational part of her mind retreated beneath a wave of desperate hope because what if? What if, after all these years of trying to save everyone else, she could finally save the one person who mattered most?

Her fingers trembled against the doorknob. The metal felt shockingly cold against her skin, anchoring her in reality for one last moment of clarity. This wasn't possible. This couldn't be real.

But from the hallway, soft and achingly familiar: "Please, Chelleybean. Let me in."

The doorknob turned under her fingers, not by her own volition but as if someone on the other side was pulling it open. Chelle stumbled back, her wet feet slipping on the floorboards as the door swung inward with excruciating slowness.

The hallway beyond was impossibly dark, a darkness that seemed to have substance and weight, like ink poured through her flat. A silhouette stood there—a woman's shape, familiar in its outline. Katherine's outline.

"There you are," the voice said, perfect in its imitation of her sister. "I've missed you so much."

The figure took a step forward, and as the dim light from Chelle's bedroom touched its face, she saw that it wasn't a face at all. Where features should have been was only smooth, blank skin stretched over bone, interrupted by a mouth that opened too wide, stretching from ear to ear.

"Don't you recognize me?" The voice emanated from that impossible mouth, the words rippling out like they were being pushed through water. "It's me, Chelleybean."

Chelle scrambled backward until her spine hit the desk. Her fingers found the TARDIS phone and she clutched it like a weapon.

"You're not Katherine," she whispered, her voice trembling as the thing in her doorway took another step forward.

The faceless head tilted, an eerily human gesture. "No," it agreed, Katherine's voice melting away into something ancient and hungry. "But I can wear her for you, if that's what you need."

As it spoke, the blank flesh where a face should be began to ripple and distort. Beneath the skin, something was moving, pressing outward as if desperate to emerge. The smooth surface bulged and stretched, tearing in places to reveal glimpses of what lay beneath—not muscle or bone, but something that gleamed wetly in the dim light, something with too many eyes.

The creature lurched forward suddenly, its movements no longer human. Its body seemed to elongate, limbs stretching impossibly as it crossed the distance between them in a single bound. The skin of its face split entirely, peeling back like petals to reveal a writhing mass of appendages surrounding a circular maw lined with concentric rings of teeth.

The last thing Chelle saw before it engulfed her was her sister's face forming briefly among the churning mass —Katherine's eyes meeting hers with an expression of pure horror, a silent scream frozen on lips that mouthed two final words:

I'm sorry.

Westlake High,
Halloween 1984

"That wedding dress in the corner? It belonged to Sienna Morrison. Or it will. Time gets slippery around certain objects in my shop.

"Sienna just wanted a killer Madonna costume for her high school reunion. What she got was a oneway ticket to Westlake High, Halloween 1984—the night of the infamous massacre that left thirteen students dead.

"Funny how the perfect outfit can change every-thing. One minute you're worried about impressing old

classmates, the next you're running from a psychopath with a machete. Makes those reunion jitters seem quaint, doesn't it?

"Some say clothes make the woman. In Sienna's case, they unmade her present and rewrote her past. But changing history has a price, and time doesn't appreciate being tampered with.

"Care to try it on? I offer excellent terms. Just don't mind the bloodstains—they always seem to come back, no matter how many times I wash them out."

Sienna Morrison slammed her palm against the steering wheel, checking the third location off her list with an aggressive stroke of her pen. Three Spirit Halloween stores, and not one had anything remotely acceptable for her reunion. Everything was packaged polyester garbage—the kind of costume that screamed "I stopped at a strip mall on my way here."

"Ridiculous," she muttered, sliding her notebook into her purse.

The sun was setting, casting long shadows across the parking lot as she stepped out of her car. Two days until the reunion, and she still didn't have a costume for the

80s-themed Halloween event. Her watch showed 6:45 PM—most stores would be closing soon.

Sienna walked briskly past her usual dry cleaner, mentally cycling through backup options. Maybe she could cobble together something from her own closet, though nothing she owned screamed "1980s Madonna" the way she wanted. She was so lost in thought that she almost missed it—a storefront that definitely hadn't been there during her last visit. Where vacant retail space had stood for months, now hung a weathered wooden sign: "SECOND CHANCES THRIFT."

She pushed open the door, and a small brass bell announced her arrival with a gentle chime. Inside, the air hung heavy with the distinct scent of old fabric and furniture polish. Unlike most thrift stores with their chaotic merchandise mountains, this place was meticulously organized. Racks of vintage clothing lined the walls in chromatic order. Glass display cases housed jewelry and accessories, each item carefully arranged as if in a museum exhibit rather than a secondhand shop.

Most surprising was the lighting—dim but intentional, like an upscale boutique rather than a thrift store. Sienna found herself stepping further inside, drawn by the unexpected atmosphere.

A string of beads clacked against one another as an elderly shopkeeper emerged from the back of the store.

The shopkeeper was thin with sharp features, moving with surprising grace despite apparent age. The eyes that settled on Sienna seemed to see more than they should.

"Looking for something specific?" The shopkeeper's voice was raspy but not unpleasant, carrying the worn texture of decades of use.

Sienna hesitated, suddenly aware of how quiet the shop was. No music, no other customers, just the faint tick of an antique clock somewhere in the back.

"I need an 80s costume for my high school reunion," she explained, regaining her composure. "Something unique, not the cheap plastic stuff they sell everywhere else."

The shopkeeper nodded knowingly, thin lips curving into a slight smile. A gesture with a bony hand directed Sienna toward a rack near the back wall. "The 80s are making a comeback. You might find something to your liking."

As Sienna made her way to the rack, she noticed her footsteps were absorbed by the thick carpet, creating an almost underwater sensation of muted sound. The vintage clothing was organized by decade, with the 80s section prominently displayed. She ran her fingers over the fabrics—authentic pieces, not reproductions.

She rifled through leather jackets with oversized shoulders, neon spandex tops, and acid-washed denim.

Each piece had potential, but nothing felt right for her Madonna costume vision. Pushing the hangers aside with increasing disappointment, she reached the end of the rack.

"Nothing here," she muttered, turning to leave. That's when something white caught her eye, tucked between two heavy winter coats on an adjacent rack.

Curiosity pulled her closer. She pushed the coats aside and found herself staring at a white lace wedding dress. The intricate pattern, the structured bodice, the layered tulle—it was unmistakable.

"This looks just like Madonna's dress from the 1984 VMAs," she breathed, carefully removing it from the rack.

The shopkeeper appeared beside Sienna with startling suddenness, moving without a sound across the carpeted floor. Those piercing eyes studied both Sienna and the dress with cryptic interest.

"Interesting choice," the shopkeeper said. "That piece has been here for years. No one's shown interest in it before."

She held the dress against herself, turning toward a nearby standing mirror. The white lace contrasted beautifully with her brunette hair and olive skin. Even unmodified, it had potential.

"Is there a changing room?" she asked.

The shopkeeper gestured toward a small curtained alcove in the corner. Sienna took the dress and slipped behind the heavy velvet curtain. The changing space was surprisingly spacious inside, with a small stool and hooks on the wall. A single bulb overhead cast a warm glow that made the white lace seem to shimmer.

She carefully stepped into the dress, half-expecting it to be too small or too large—thrift store sizes were always a gamble. But as she pulled it up and fastened the back, Sienna was startled by how perfectly it fit, as if it had been tailored specifically for her body.

Sienna turned slowly, examining how the fabric hugged her curves in all the right places. Even without the modifications she'd planned—the "Boy Toy" belt, the crucifix necklace, the fingerless lace gloves—the resemblance to Madonna's iconic look was undeniable. The structured bodice pushed her up in the exact right way, and the layered skirt fell at precisely the right length.

"This is . . . perfect," she whispered to herself.

She stepped out from behind the curtain to find the shopkeeper waiting, hands clasped patiently before them.

"I'll take it," Sienna decided, reaching for her purse. As the shopkeeper moved to the old-fashioned register at the front counter, Sienna followed, suddenly curious. "Do you know where it came from originally? It's in such good condition."

The brass keys of the ancient register clicked loudly in the quiet shop as the shopkeeper rang up the sale. "Local donation, that's all I know. Been sitting here longer than I can remember." The shopkeeper's movements were efficient despite age-spotted hands, carefully folding the dress into a black garment bag that appeared from beneath the counter.

Sienna handed over her credit card, watching as it was swiped through an outdated card reader. When the receipt printed, the shopkeeper slid it across the counter with a pen and Sienna signed it.

The shopkeeper handed over the garment bag with both hands, as if passing along something of great importance. "Enjoy your reunion."

"Thank you," Sienna replied, draping the bag carefully over her arm. "You've saved me from showing up in some tacky costume. My classmates would never have let me live it down."

She turned and walked to the door, the brass bell chiming again as she pushed it open. Outside, the evening had grown darker, the strip mall parking lot now illuminated by buzzing fluorescent lights.

Behind her, unseen, the shopkeeper watched through the window, perfectly still. The neon "OPEN" sign in the window flickered once, twice, then stabilized—but not before momentarily revealing different lettering underneath, too quick to read.

SIENNA'S LIVING room was a mess. Her coffee table had vanished beneath piles of fabric scraps, thread, scissors, and craft glue. The wedding dress hung from her mannequin in the corner—a purchase she'd made years ago during an ambitious phase when she thought she might start making her own clothes. She'd used it twice.

On the wall, she'd taped printouts of Madonna at the 1984 VMAs. She'd grabbed screenshots from every angle she could find online. Madonna standing defiantly, the white dress contrasting with that bold "BOY TOY" belt. It was perfect for the reunion—something no one else would think to do.

"Alright," Sienna muttered, picking up her measuring tape. "Let's not screw this up."

She measured twice for the belt, marking the faux leather with a white pencil before cutting. Her first attempt at stenciling "BOY TOY" across it went crooked, forcing her to start over with her last piece of good material.

"Damn it," she said, tossing the ruined piece aside. She glanced at her phone—3:24 PM. Plenty of time still.

The afternoon wore on. Sienna attached lace to a pair of gloves she'd ordered online while watching a YouTube tutorial on her laptop. She sewed costume pearls onto the bodice, pricking her finger twice and leaving a tiny bloodstain she hoped wouldn't show. She frayed specific edges of the tulle, trying to match Madonna's "just woke up like this" vibe.

By 7:30, her back hurt and she was starving. She hadn't eaten since a granola bar at lunch. She stood up, stretching until her spine cracked, and looked over her work.

"Not bad," she told herself, "but not done either."

Sienna pulled the costume from the mannequin and held it up to herself, checking her reflection in the full-length mirror propped against her living room wall. The belt needed to sit lower. She made a mental note, then grabbed a notepad from the coffee table, pushing aside scraps of fabric to find a pen.

"Lower belt," she scribbled. "Add more lace to right sleeve. Find bigger crucifix?"

Her stomach growled, reminding her that creative work was still work. She ordered Thai food from the place down the street, then returned to the dress while waiting for delivery. The costume was taking shape, but it still needed something to make it pop—that distinctive Madonna edge that separated iconic from just another 80s throwback.

After wolfing down pad thai at her kitchen counter—carefully away from the costume—Sienna worked until nearly midnight, finally collapsing into bed with glue on her fingers and a satisfied exhaustion.

The next day flew by in a blur of last-minute adjustments. She found a larger crucifix at a costume jewelry shop during her lunch break, and by evening, the costume was finally complete—hanging ready for the reunion tomorrow night.

THE NIGHT of the reunion arrived with a mix of excitement and anxiety that Sienna hadn't expected. Her bathroom counter was cluttered with more products than she'd used in the last month. Mascara, three shades of eyeshadow, red lipstick, foundation, powder, and hairspray—lots of hairspray. Madonna's "Like a Virgin" played from her portable speaker, setting the mood.

"You're really committing to this," she told her reflection as she applied a heavy layer of blue eyeshadow.

It felt ridiculous, but that was the point—1980s excess in all its glory.

She carefully traced a beauty mark above her upper lip with eyeliner, then stepped back to check the effect. The makeup alone was transformative, but her brown hair was still all wrong.

Sienna hadn't gone blonde since a disastrous experiment in college, but tonight required the full transformation. She'd bought a temporary color spray that promised to wash out in one shampoo. She sectioned her hair with clips and began spraying, the chemical smell making her eyes water despite the bathroom fan running at full blast.

"This better work," she muttered, turning her head side to side as the brown disappeared under layers of golden blonde.

Once her hair was fully colored, she attacked it with a teasing comb, backcombing sections until they stood up in wild disarray. She sprayed each section with hairspray, the aerosol creating a thin mist that settled on everything in the bathroom. Her fingers felt sticky, and she was pretty sure she'd inhaled enough chemicals to concern an EPA inspector.

The final step was scrunching the teased sections with her hands, creating that deliberate mess that had somehow been the height of fashion. One last coat of hairspray locked it all in place.

Sienna stepped back, taking in the full effect. The wild blonde hair, the dramatic makeup, the white lace dress with the "BOY TOY" belt sitting provocatively on her hips, and the crucifix dangling between her breasts—the transformation was startling.

"Holy shit," she whispered. For a moment, it wasn't Sienna Morrison in the mirror but Madonna herself, circa 1984. She turned side to side, watching how the dress moved with her. All those hours of work had paid off.

She grabbed her phone from the counter and took several selfies, trying different poses—hand on hip, crucifix held up to camera, a mock-innocent pout. She scrolled through them, selected the best one, and posted it to Instagram with the caption: "Ready for my reunion tour. #LikeAVirgin #Class2004 #ReunionNight"

The likes started coming in immediately. A text from Jordan followed seconds later:

OMG perfect!! See you there!

Sienna checked the time on her phone—7:38 PM. The reunion started at eight, and she wanted to make an entrance, not be the first one there. She gathered her essentials—phone, lip gloss for touchups, ID, and credit card—into a small purse with a thin strap that wouldn't interfere with the costume.

"Keys," she reminded herself, grabbing them from the hook by the door.

She took one final look in the hallway mirror, adjusting the crucifix necklace so it hung perfectly centered. The weight of it felt strange against her chest, heavier than it looked. She touched it briefly, feeling an odd moment of déjà vu that she couldn't place.

"Get it together, Si," she told herself. "It's just a high school reunion, not the actual VMAs."

Sienna locked her apartment door behind her, the excitement of the night ahead pushing away the momentary unease. The costume was perfect, she looked amazing, and for once, she wasn't showing up at a high school event feeling like she hadn't quite gotten it right. Tonight was different. Tonight she would be remembered.

As she walked to her car, a neighbor from across the hall spotted her and did a double-take.

"Whoa! Madonna lives!" he called out, giving her a thumbs up.

Sienna grinned and struck a pose, confidence surging. If the reaction at the reunion was half as good, the hours of work would be worth it.

She slid into her car, careful not to catch the dress on anything, and checked her appearance one more time in the rearview mirror. The woman looking back at her seemed different somehow—bolder, ready for whatever the night might bring.

SIENNA CRUISED through town with her windows cracked despite the October chill. Her specially curated 80s playlist blasted from her car speakers, and she sang along with abandon to Cyndi Lauper's "Girls Just Want to Have Fun." At a red light, she caught another driver watching her mid-performance and just grinned, too caught up in the moment to feel embarrassed.

"I come home in the morning light," she belted out, drumming her hands on the steering wheel. "My mother says, 'When you gonna live your life right?'"

The streetlights flickered on as dusk settled over Westlake, casting intermittent spotlights that illuminated her excitement. She hadn't expected to feel this way about the reunion. For years, high school had been a memory she'd mostly shoved aside—not terrible, but not worth revisiting. Yet here she was, practically giddy at the prospect of seeing people she'd once counted down the days to escape.

Westlake High appeared ahead, its red-brick exterior outlined with orange Halloween lights. The parking lot was packed, forcing Sienna to circle twice before finding a spot at the far end. She switched off the engine mid-

Madonna chorus and sat for a moment, suddenly aware of her rapid heartbeat.

"You've got this," she whispered, adjusting the costume in her rearview mirror. She reapplied her red lipstick, touched up the beauty mark, and fluffed her teased blonde hair.

Taking a deep breath, she stepped out of the car, immediately wobbling on her white heels as they clacked against the asphalt. The evening air hit her exposed skin with a sharpness that made her gasp. Fallen leaves crunched beneath her feet as she made her way toward the entrance, their earthy scent mixing with the distant smell of pumpkin spice and artificial fog.

The path to the school entrance was lined with carved jack-o-lanterns, their candles flickering in the evening breeze. Fake cobwebs stretched across bushes, and a large banner hung over the main doors: "Westlake High 20 Year Reunion Halloween Bash." Seeing the school name in big bold letters made the whole thing suddenly real in a way online invitations hadn't.

A couple dressed as Sonny and Cher walked several paces ahead of Sienna, the woman's long straight hair swinging dramatically with each step. At the sound of Sienna's heels on the pavement, they turned to look. The woman's eyes widened, taking in the Madonna costume with clear appreciation. She nudged her partner and gave Sienna an approving nod.

"That is commitment," she called back. "You're gonna win the costume contest for sure."

"Thanks," Sienna replied, a small burst of pride warming her chest. The hours of work were already paying off.

Her phone buzzed in her purse. She pulled it out to find a stream of texts from Jordan:

Where are you???

You better be wearing that Madonna outfit

Tina is here and she's BALD now. Not like, shaved. Like, BALD.

Eric brought his 25-year-old girlfriend.

We're all dying.

Sienna snorted, typing back a quick "Almost there" before tucking her phone away.

The bass of "Wake Me Up Before You Go-Go" thumped through the walls as she approached the check-in table. A woman in a "Flashdance" costume—complete with leg warmers and off-the-shoulder sweatshirt—sat behind it, surrounded by name badges organized alphabetically.

"Hi," Sienna said, fishing in her purse for her ticket. "Sienna Morrison."

"Oh my god, Sienna? I wouldn't have recognized you

with the blonde hair!" The woman looked up with a huge grin. "It's me, Becky Turner! Well, Becky Whittaker now."

Sienna squinted, trying to place her. Becky from AP Chemistry? Debate team Becky? She couldn't remember, but smiled broadly anyway.

"Becky! You look great!" she exclaimed, hoping her enthusiasm would mask her lapse in memory.

Becky beamed, clearly pleased with the recognition. "So do you! That Madonna costume is amazing. Did you make it yourself?"

"Most of it, yeah," Sienna admitted. "Found the wedding dress at a thrift shop and modified it."

"Talented as always," Becky said, thumbing through a stack of badges. "You were always creative—remember that mural project in Mr. Davidson's class?"

Sienna nodded vaguely, having no recollection of any mural project. Twenty years had washed away so many details.

"Here you go," Becky handed over a plastic name badge. "Sienna Morrison—Class of 2004" was printed in bold black letters against a background of the school's blue and gold colors.

"Thanks," Sienna said, carefully pinning the badge to her dress where it wouldn't ruin the look too much. "Who else is here so far?"

"Pretty much everyone," Becky replied. "Jordan's been asking about you. I think he's by the punch bowl—which, by the way, is already spiked. Some things never change."

Sienna laughed. "Good to know." She straightened her shoulders, adjusted her crucifix one last time, and walked toward the gymnasium doors, from which music and laughter spilled out into the hallway.

THE HEAVY GYMNASIUM doors swung open under Sienna's push, releasing a wave of sound that washed over her. Cyndi Lauper's "Time After Time" played at full volume, the familiar melody unexpectedly triggering a wave of nostalgia.

"If you're lost, you can look, and you will find me..."

Sienna stepped inside, momentarily dazzled by the scene. Crepe paper streamers in blue and gold hung from the ceiling in elaborate crisscross patterns. Helium balloons floated in clusters at each corner of the room. A disco ball spun lazily overhead, sending fragments of light spinning across the floor, walls, and faces of the dancers.

She paused at the threshold, her smile of anticipation slowly fading. Sienna scanned the crowd, expecting to see former classmates, now in their late thirties, dressed in 80s costumes. Instead, she saw only teenagers—fresh-faced, awkward, radiating that distinctive high school energy she remembered so well. They danced and laughed, oblivious to her confusion.

"What the hell?" she muttered, stepping further into the gym.

She looked more closely at the banner hung across the far wall, expecting to see "Welcome Back Class of 2004." Instead, in hand-painted letters, it read: "Halloween Dance 1984."

Sienna blinked hard, certain she'd misread it. But the banner remained unchanged: "Halloween Dance 1984."

"This can't be right," she whispered. She instinctively reached for her purse to call Jordan, fingers searching for her phone that she'd just used in the parking lot minutes ago. Her stomach dropped when she found only lip gloss, her ID, and a few dollars. No smartphone.

A creeping sense of unreality washed over her as details she'd initially missed suddenly came into focus— the girls' puffy sleeves and high-waisted jeans, the boys' mullets and Members Only jackets. The music wasn't a retro playlist; it was current. The decorations weren't nostalgic throwbacks; they were simply what was available at the time.

"This isn't happening," Sienna muttered, backing toward the door she'd just entered. Her heel caught on something, nearly causing her to stumble. A boy with a mullet grabbed her elbow to steady her.

"Whoa, careful there," he said, then moved on without a second glance.

Sienna pushed back through the gym doors into the hallway, half-expecting to find herself back in 2024. But the hallway remained unchanged—trophy cases displaying achievements from decades past, bulletin boards with hand-drawn posters advertising clubs and events, all dated 1984.

"Becky?" she called out, hurrying toward the check-in table. But Becky was gone. In her place sat a middle-aged woman Sienna didn't recognize, collecting tickets from a group of teenagers.

"Excuse me," Sienna approached her, trying to keep her voice steady. "I'm looking for Becky Turner?"

The woman looked confused. "Turner? I don't think we have a Becky Turner working check-in tonight."

Sienna backed away, her heart hammering in her chest. She turned and ran toward the main entrance, her heels clicking frantically on the linoleum. She pushed against the heavy doors, but they wouldn't budge. Locked.

She tried another exit with the same result. Trapped.

"This is a dream," she said aloud, pressing her hands against her temples. "This is a stress dream about the reunion, and I'm going to wake up any minute."

She pinched her arm hard enough to leave a mark. Nothing changed.

She caught her reflection in a trophy case glass—blonde hair, Madonna costume, wide, frightened eyes. Behind her reflection, a wall calendar clearly showed October 1984, with Halloween circled in red marker. Beneath it hung a recent homecoming photo featuring hairstyles and fashions unmistakably from that era.

She walked slowly back toward the gymnasium, drawn by the music and the relative safety of a crowd. She watched the teenagers laughing and dancing, living in a moment that for her had passed four decades ago. Whatever was happening, she needed to stay calm and figure out her next move.

A sudden wave of dizziness washed over Sienna, forcing her to sit on the bottom row of the bleachers. The music, the lights, the unfamiliar-yet-familiar surroundings —it was all too overwhelming. She rubbed her temples, trying to calm the spinning sensation.

"Are you okay?"

Sienna looked up to find a teacher standing over her —a man in his fifties wearing a brown tweed jacket with elbow patches.

"I'm . . . fine," she managed. "Just a little overwhelmed."

"New school dances can be intimidating," he said kindly. "But everyone's welcome at Westlake, even visitors." He nodded at her costume. "Great Madonna outfit, by the way. Very current."

"Thank you," she replied automatically, watching as he moved away to break up a cluster of boys who were roughhousing near the punch bowl.

She scanned the room, trying to focus on the immediate. If she was stuck here, even temporarily, she needed to blend in. The last thing she wanted was to draw too much attention to herself. Who knew what consequences that might have?

A flash of white across the gym caught her attention —another Madonna costume, worn by a blonde girl surrounded by friends. The girl laughed at something, tossing her head back in a way that spoke of confidence and ease, the center of her social circle. Something about her seemed vaguely familiar, though Sienna couldn't place her.

"Focus," she told herself. "First things first. Figure out where you are, then worry about how to get back."

Sienna turned her attention to the decorations on the wall behind her. A corkboard covered with photographs from what must have been recent school events. The fashions, the hairstyles, the film grain quality of the images—all undeniably from the 1980s.

She felt a strange disconnect watching these teenagers, knowing what lay ahead for them—the fall of the Berlin Wall, the dawn of the internet, 9/11, smartphones. An entire future these kids couldn't imagine, all commonplace history to her.

"What am I supposed to do here?" she wondered.

This was obviously some bizarre hallucination or vivid dream—it had to be. Eventually, she'd wake up or snap out of it and find herself back in 2024.

But what if she didn't?

Sienna straightened her shoulders, making a decision. Until she figured out how to get back to her own time, she'd play along. She'd blend in as much as possible and avoid doing anything that might . . . what? Change the future? Was that even possible?

She pushed herself up from the bleachers, smoothing the white lace of her costume. If she was going to be stuck at a high school dance in 1984, she might as well not spend it huddled in a corner having an existential crisis. At minimum, she needed water and a moment to collect herself.

Sienna made her way toward the refreshment table, careful not to make eye contact with too many people. The punch bowl was surrounded by students, laughing and filling plastic cups with the bright red liquid that certainly contained more than just fruit juice, based on their behavior.

She spotted a stack of cups next to a cooler marked "WATER" and headed there instead. As she reached for a cup, her hand collided with someone else's.

"Sorry," Sienna said automatically, looking up into the face of a girl with bright blue eyeshadow and teased blonde hair. The other Madonna.

"No problem," the girl replied with a friendly smile. She studied Sienna's costume with undisguised admiration. "Wow, your Madonna is perfect. Like, scary perfect. Did your mom help you make it?"

"I, uh, made it myself," she answered, filling her cup with water to hide her discomfort.

"It's incredible," the girl continued, her eyes lingering on the details of the costume. "The 'Boy Toy' belt looks exactly like the real thing." She extended her hand. "I'm

Heather, by the way. Heather Morgan. Senior. I don't think I've seen you around before."

"Sienna," she replied, shaking Heather's hand. "I'm . . . visiting."

"Cool! Where from?" Heather asked, taking a sip from her punch cup. The smell of cheap vodka wafted from it.

"Just . . . . nearby," Sienna hedged, searching for a plausible explanation. "I'm staying with relatives for a few days."

Heather nodded, accepting this vague explanation without question. At seventeen, other people's backstories weren't particularly interesting unless they directly affected her social standing.

"Well, whoever your relatives are, they picked a good weekend for you to visit. Our Halloween dance is always epic." She gestured around the gymnasium. "Last year, Robbie Martinez spiked the punch so bad that Principal Davis passed out in his office." She laughed at the memory, then frowned slightly. "Wait, who are you here with?"

"No one," Sienna admitted. "My . . . cousin was supposed to come, but got sick at the last minute." The lie felt clunky on her tongue, but Heather didn't seem to notice.

"That's a bummer," Heather said. "No one should be

alone at a dance, especially not another Madonna." She brightened, as if struck by a brilliant idea. "Hey, come meet my friends! They'll love your costume."

Before Sienna could respond, Heather had linked arms with her, leading her across the dance floor toward a group of well-dressed teenagers lounging near the bleachers. Everything about them screamed "popular crowd"— their confident postures, their expensive clothes, the way other students glanced at them with a mixture of envy and admiration.

"Guys," Heather announced, "this is Sienna. She's visiting and came to the dance alone, so I've officially adopted her for the night."

Sienna found herself facing four teenagers who studied her with varying degrees of interest. A blonde boy in a Michael Jackson "Thriller" jacket looked her up and down with obvious appreciation. A girl in a Ghostbusters jumpsuit gave a friendly wave.

"This is Brad, Melissa, Jason, and Kim," Heather said, pointing to each in turn. "Sienna has the best Madonna costume I've seen. Even better than mine!"

"Nice to meet you," Sienna said, suddenly aware of how strange it was to be introduced to a group of teenagers as if she were one of them.

"Where'd you get the costume?" Kim asked, examining the details. "The lace looks vintage."

"I found the dress at a thrift shop," Sienna explained, sticking as close to the truth as possible. "Then added the accessories myself."

"Thrift shops have all the best stuff," Melissa agreed. "My mom never lets me go to them though. She says they're full of 'other people's problems.'" She rolled her eyes dramatically.

Before Sienna could formulate a response, a teacher approached their group, a camera hanging around his neck. He had thick glasses and a cardigan that screamed "yearbook advisor."

"Costume photos for the yearbook," he announced, gesturing toward a Halloween-decorated backdrop near the DJ booth. "How about the popular crew and their new friend?"

"Come on!" Heather tugged at Sienna's arm. "We need to document the Madonna twins!"

"I'm not sure—" Sienna began, but Heather was already pulling her toward the photo area.

"It'll be fun," Heather insisted.

The group arranged themselves in front of the back-drop—a black paper sheet with orange and white paper ghosts and pumpkins taped to it. Heather positioned Sienna right next to her in the center.

"Madonna squared," she joked, striking a pose with one hand on her hip.

"Smile everyone!" the teacher said, adjusting his camera lens. "This is going in the yearbook!"

The thought sent a chill through Sienna. The yearbook. Historical documentation. Her—a woman from 2024—appearing in a 1984 high school yearbook photo. What would that mean for the timeline? Would her younger self someday see this photo? The implications made her head spin.

Before she could protest, the flash went off, momentarily blinding her. Sienna blinked rapidly, spots dancing before her eyes.

"One more!" the teacher called out. Heather threw her arm around Sienna's shoulders, pulling her closer. The flash popped again.

"That's great," the teacher said, scribbling something in a small notebook. "Names?"

The students called out their names in succession. When it came to Sienna, Heather jumped in, "And this is Sienna..." She paused, looking to Sienna expectantly.

"Morrison," Sienna supplied automatically, her mouth working before her brain could catch up.

The teacher nodded, noting it down. "Thanks, everyone. These will be great for the Halloween spread."

As the group dispersed, Jason laughed and called back, "Hopefully we all look good. These pictures are gonna be around forever!"

"You okay?" Heather asked, noticing Sienna's expression. "You look like you've seen a ghost."

Sienna forced a smile. "Just a little overwhelmed. I don't know many people here."

"Come on, let's get some more punch," Heather said, taking Sienna's arm. "The night's just getting started!"

Heather led Sienna back toward the refreshment table, weaving through dancing couples as Michael Jackson's "Thriller" began playing. The crowd cheered, several students immediately breaking into the iconic zombie dance from the music video.

Heather ladled punch into a fresh cup and handed it to Sienna. She accepted the cup but didn't drink. The sweet smell of cheap vodka mixed with even cheaper fruit juice wafted up to her. Her teenage self might have welcomed it, but right now, she needed all her wits about her.

As they talked, Sienna studied Heather more carefully. There was something vaguely familiar about her— not in the way of someone Sienna had known personally, but more like a face from an old photograph. The name Heather Morgan tugged at her memory, but she couldn't place it.

"Your costume really is amazing," Heather said, returning to their earlier conversation. "My mom helped me with mine. She thinks Madonna is going straight to hell, but she still helped with the sewing."

Sienna smiled despite herself. "Your mom sounds interesting."

"That's one way to put it," Heather laughed. "Hey, want to check out the haunted hallway? The drama club set it up in the English wing, and it's supposed to be actually scary this year."

"Sure," Sienna agreed, seeing an opportunity to explore more of the school and maybe find clues about what had happened to her. Plus, staying with Heather seemed safer than wandering alone.

As they walked out of the gymnasium, Sienna felt a strange sense of protectiveness toward the teenager beside her. Heather couldn't be more than seventeen or eighteen, with all the confidence and vulnerability that age entailed. There was an innocence to her that made Sienna ache with a nostalgia she hadn't expected to feel.

"So when did you decide to do Madonna?" Sienna asked as they walked down the hallway, their heels clicking against the linoleum.

"The second I saw her on MTV," Heather said enthusiastically. "That VMA performance was just so . . . I don't know, brave? My dad nearly had a heart attack when he saw it. Said she was corrupting America's youth." She rolled her eyes. "Parents just don't get it."

Sienna bit back a smile at the quintessentially teenage sentiment. "Mine were the same way," she said, which was true. Her own parents had been conservative and cautious about pop culture influences, something she'd rebelled against in her own adolescence.

They rounded a corner, and the sound of screams—playful ones—echoed from the English wing ahead. Black paper covered the windows of classroom doors, and fake cobwebs hung from the ceiling. A sign read "ENTER IF YOU DARE" in dripping red letters.

"Drama club goes all out every year," Heather explained. "Last year, Jason Mitchell had to go home because he peed himself a little."

Two drama students in ghoulish makeup stood at the entrance, collecting tickets for the haunted hallway.

"Two," Heather said, digging in her purse for the small orange tickets.

Sienna touched her arm. "I don't have a ticket."

"It's cool," Heather handed the boy two tickets. "I've got extras. Brad always buys a bunch for the group."

As they entered the darkened hallway, eerie music played from hidden speakers. The ceiling lights had been covered with red and purple gels, casting the corridor in an otherworldly glow. Drama students in various monster costumes jumped out from classroom doorways, eliciting screams from the students ahead of them.

"This is actually pretty good," Sienna admitted, impressed by the production value achieved with clearly limited resources. A Frankenstein's monster lurched toward them from an alcove, arms outstretched.

Heather shrieked and grabbed Sienna's arm, then immediately laughed at herself. "I knew he was going to be there and still jumped!"

Sienna found herself smiling. There was something refreshing about Heather's unfiltered reactions—so different from the curated responses of people her own age in 2024, always aware of how they might be perceived.

As they moved deeper into the haunted hallway, Sienna's thoughts drifted back to her situation. What if she couldn't get back? What if she was truly trapped in 1984? The thought sent a wave of panic through her, which she fought to suppress.

"You look worried," Heather said, noticing Sienna's expression in a flash of strobe light. "Are the monsters too scary?" Her tone was gently teasing.

"No, just . . . thinking about something else," Sienna replied.

"Let me guess—boy problems?" Heather nudged her playfully. "That's usually what puts that look on my friends' faces."

Sienna couldn't help but laugh. "Not exactly."

"Well, whatever it is, forget it for tonight," Heather advised with the sage wisdom of a teenager who believed all problems could be solved by simply ignoring them. "That's what Halloween is for—pretending to be someone else for a night."

The irony wasn't lost on Sienna.

A zombie jumped out at them, and Heather screamed again, clutching Sienna's arm tighter. As they passed through the final section, a strobe light pulsed rapidly, making the motion of fleeing students ahead of them appear jerky and disconnected.

They emerged from the haunted hallway back into the regular corridor, Heather still giggling from the final scare. The normal fluorescent lighting seemed painfully bright after the atmospheric darkness they'd left behind.

"That was actually pretty decent," Sienna said, blinking to adjust her eyes.

"Drama club's been working on it for weeks," Heather replied. She checked a small watch on her wrist —a genuine Swatch, Sienna noted, not a retro reproduction. "We should head back to the gym. They're announcing the costume contest winners soon, and I think we both have a shot."

As they walked back toward the gymnasium, Sienna noticed how comfortably Heather had adopted her, bringing her into her circle without hesitation. There was a genuine warmth there that contradicted every stereotype about popular high school girls.

"Can I ask you something?" Sienna said suddenly.

"Sure."

"Why are you being so nice to me? I'm a complete stranger."

Heather looked surprised by the question. "Why wouldn't I be? We're wearing the same costume, for one thing. That makes us automatic friends." She shrugged. "Besides, you seemed lost, and nobody should be alone at a dance."

The simple, earnest response caught Sienna off guard. There was no calculation, no social media-trained performance of kindness for external validation—just a spontaneous act of inclusion that seemed to come naturally to Heather.

They approached the gymnasium doors, where the sound of music pulsed steadily. But as they reached for the handle, the music abruptly cut out, leaving an unnatural silence in its wake.

"What happened to the music?" Heather frowned.

Sienna felt a sudden, inexplicable chill. Something about the silence felt wrong—heavy and anticipatory, like the air pressure dropping before a storm.

They pushed through the double doors into the gymnasium, where confused students were looking around and murmuring to each other. The DJ was fiddling with his equipment, checking connections and muttering to himself.

The gymnasium lights flickered once, twice, then stabilized. A ripple of nervous laughter went through the crowd.

"Halloween spooks!" someone called out, eliciting a few more laughs, though they sounded forced.

Heather rejoined her friends, pulling Sienna along with her. "What's going on?" she asked Brad, who was watching the DJ with mild interest.

"Equipment failure," he said with a shrug. "Or maybe the classic 'circuit breaker blown by too many decorations' scenario."

A distant metallic clanking sound echoed through the gym, cutting through the murmurs of conversation. It was rhythmic, deliberate—like chains being pulled taut. The sound sent a jolt of recognition through Sienna, though she couldn't place why.

"What was that?" Heather asked, her previous buoyancy fading.

"Probably just part of the haunted house setup," Brad suggested, though he didn't sound convinced.

Mr. Henley walked toward the main doors, his brow furrowed with concern. "I'll check it out," he called to another teacher. "Everyone stay here and enjoy the party."

As if on cue, the DJ got the music working again, starting up The Police's "Every Breath You Take." Students gradually returned to dancing, though an undercurrent of unease remained.

Sienna watched Mr. Henley push through the gymnasium doors, a cold knot forming in her stomach. Something about him walking toward those doors triggered a distant memory—something about Westlake's history, a story she'd heard long ago.

Sienna tried to grasp at the memory teasing the edges of her mind. Something bad had happened at Westlake, something everyone knew about but rarely discussed. Something on Halloween.

"Hey," Heather nudged her, "you look like you've seen a ghost. Again."

"I just have a weird feeling," Sienna admitted, still watching the door Mr. Henley had disappeared through.

"About what?"

Before Sienna could answer, the music cut out again. This time, the lights dimmed for several seconds before returning to full brightness. The disjointed technical issues were beginning to feel deliberate rather than accidental.

Brad sighed dramatically. "If they can't keep the power on, they should just cancel the dance and give us our money back."

"Don't be such a downer," Melissa told him. "It's Halloween. Weird stuff is supposed to happen."

Sienna's attention remained fixed on the gym doors. Mr. Henley had been gone for several minutes now. Shouldn't he have returned if it was just a simple issue to check on?

"I'm going to see what's happening," she said, starting toward the doors.

Heather caught her arm. "Mr. Henley said to stay here."

"I just want to look," Sienna assured her. "I'll be right back."

She had taken only a few steps when the gymnasium doors swung open. Mr. Henley stood in the doorway, his body silhouetted against the dimmer hallway lights. There was something off about his posture—too rigid, too still.

For a heartbeat, the gym was utterly silent. Then Mr. Henley stumbled forward, revealing a dark, glistening stain spreading across the front of his shirt. He opened his mouth as if to speak, but only a wet, gurgling sound emerged.

Behind him, a figure appeared in the doorway—tall, wearing a clown mask, holding something that caught the light with a sick gleam.

Mr. Henley's hands rose to his throat in a jerky

motion, his expression one of disbelief rather than pain. His fingers came away red, glistening under the gymnasium lights. He took two more stumbling steps into the room before collapsing onto the polished floor with a dull thud.

Students nearest the door froze, their minds struggling to process what they were seeing. The bright arterial blood pooling around Mr. Henley looked almost theatrical under the colored dance lights—too vivid, too red to be real.

Someone laughed nervously. "Nice special effects, Mr. H," a boy called out. "Really realistic."

The figure in the clown mask stepped fully into the doorway. Tall, wearing a dark jacket, holding a machete dripping with fresh blood. The mask's exaggerated smile created a grotesque contrast to the violence just committed.

Mr. Henley made a horrible rattling sound, his legs kicking reflexively against the floor. His throat had been cut deeply enough to expose cartilage—a precise slash that severed his carotid artery. The growing puddle of blood made it clear this was no Halloween prank.

A girl screamed—high-pitched and primal. The sound broke the stunned silence, and chaos erupted.

Students scattered in all directions. Some ran for the

far exits, others froze in terror. The DJ abandoned his equipment, knocking over speakers in his haste to escape.

"What the hell?" Brad shouted, pulling Melissa behind him.

The masked figure moved with purpose, stepping over Mr. Henley's now-still body. He swung the machete at the nearest student—a boy in a vampire costume. The blade sliced through costume and flesh alike, opening a deep gash across the boy's chest. Blood sprayed across nearby students, triggering more screams as they stumbled backward.

"It's locked!" someone shouted from an exit door, yanking frantically on the handle. "All the doors are chained shut!"

Panic intensified as students realized they were trapped. The figure continued his steady advance into the gymnasium, machete swinging in controlled, practiced arcs. The blade caught a girl's arm as she fled, opening it to the bone. She fell screaming, clutching the wound as blood pulsed between her fingers.

Heather grabbed Sienna's arm, her fingers digging in painfully. "We need to hide," she hissed, eyes wide with terror.

Sienna's mind suddenly pieced together the horrible truth. Halloween night, 1984. A masked killer. Westlake High School. The story that had been whispered for decades in her hometown—the Halloween Massacre. She was living through an infamous tragedy that had haunted Westlake's history for generations.

"This way," Sienna said, grabbing Heather's hand and pulling her toward a side corridor as screams filled the gymnasium behind them. Survival instinct kicked in, pushing aside her shock and disbelief.

They weren't alone in their flight. Several other students had the same idea, racing down the hallway ahead of them. Sienna's heels clacked loudly on the linoleum floor, each step echoing like gunshots.

"In here!" a girl ahead of them shouted, pushing open the door to the girls' bathroom. Sienna and Heather followed, along with three other terrified students.

Inside, fluorescent lights buzzed overhead, casting harsh shadows. The bathroom was exactly as Sienna remembered from her school days—down to the faded pink tile and scratched mirrors. The familiarity made the horror even more surreal—this was her school, just decades earlier.

"What the hell is happening?" a girl in a Princess Leia costume demanded, mascara running down her cheeks. "Who was that?"

"Did you see what he did to Mr. Henley?" another student whispered, voice cracking. "He cut his throat open like it was nothing."

Heather paced anxiously. "We need to barricade the door," she said, surprising Sienna with her practicality.

The group sprang into action, dragging metal trash cans and moving a small supply shelf in front of the door. The makeshift barrier looked flimsy against the horror outside.

Muffled screams continued to echo down the hallway. Something heavy crashed against a locker, followed by more screaming. Then a wet, meaty thud.

"Shouldn't we call for help?" Princess Leia asked, looking around wildly.

"The office has a phone," a boy in a pirate costume suggested. "But it's across the school."

"Nobody's going to the office," Sienna said firmly. "That's suicide." The certainty in her voice made everyone look at her, but she couldn't explain how she knew this—that she'd heard this story before, from the other side of history.

A sudden banging on the bathroom door made everyone jump. The barricade shifted slightly with each impact.

"Let me in!" a male voice shouted, desperate and pained. "Please! He's coming!"

Sienna and Heather exchanged looks, then moved quickly to shift the barricade. They opened the door just enough for a boy in a letterman jacket to stumble through, immediately replacing the barricade after.

The boy collapsed against the sinks, breathing heavily. His chest had been sliced open diagonally, the letterman jacket's fabric parted to reveal glistening muscle and white bone beneath. Blood pulsed from the wound with each heartbeat, spreading across the white porcelain in crimson rivulets.

"Oh my god, Danny," Heather gasped, rushing to his side. "What happened?"

Danny's face was ashen, his eyes unfocused. Blood bubbled at the corner of his mouth when he spoke. "It's Teddy," he managed between labored breaths. "Teddy Kopeki. He's gone crazy."

Teddy Kopeki. The name sent a chill through Sienna. The infamous killer from Westlake's history. The Halloween Massacre of 1984. She was right about what today was.

"He's chained all the exits," Danny continued, each word clearly painful. "He's killing everyone."

Princess Leia began to sob quietly. The pirate put an arm around her shoulders, his own hands trembling visibly.

Sienna moved to Danny, examining his wound. The gash was deep, slicing through pectoral muscle and exposing a portion of his ribcage. Blood continued to seep between his fingers as he clutched his chest, the crimson liquid flowing in steady pulses that indicated an artery had been hit. His white t-shirt was now completely soaked, the fabric clinging to his skin.

"How many others like him out there?" Heather asked, trying to sound calmer than she felt.

"Just him," Danny said, coughing wetly. Blood spattered his lips and chin, droplets landing on the white sink. "But he's . . . he's everywhere. Moving like a ghost." His breathing became more labored, a rattling sound emerging from deep in his chest.

The fluorescent lights flickered overhead, casting momentary shadows that made everyone flinch.

"Did anyone call the police?" Heather asked, her voice tight with fear.

"Tried," Danny whispered. "He cut the phone lines." His voice trailed off as his eyes began to lose focus. The hand pressed to his chest slowly slackened.

"Danny?" Heather shook him gently, then more urgently. "Danny!"

But Danny's gaze had turned glassy and vacant. His chest no longer rose and fell with breath. A final trickle of blood ran from the corner of his mouth, sliding down his chin to join the growing pool on the floor.

For a moment, the bathroom was completely silent except for the steady drip of Danny's blood hitting the tile floor. Each drop echoed in the quiet—a metronome counting seconds of borrowed time.

"Oh my God," Princess Leia whispered, pressing her hands to her mouth. "He's dead. He's actually dead."

Sienna stood, wiping Danny's blood from her hands onto her once-pristine white dress. The bright red stains spread across the lace like watercolor on paper. She'd never seen someone die before—not like this, not right in front of her.

"We can't stay here," she said, her voice steadier than she felt. "If Teddy is working his way through the school, he'll find us eventually."

"Are you crazy?" the pirate hissed, eyes wide with disbelief. "Out there is certain death. In here we at least have a chance!"

Heather looked between them, then at Danny's body. Blood continued to pool beneath him, creeping across the white tile in a widening stain.

"Maybe Sienna's right," she said, her voice small but determined. "We're sitting ducks in here."

Princess Leia shook her head violently, mascara streaking her cheeks. "I'm not going out there. You see what he did to Danny. To Mr. Henley." She gestured toward the door. "He's butchering people!"

Sienna paced the bathroom, mind racing. The Halloween Massacre of 1984—thirteen students died and scores more severely injured before Teddy Kopeki was finally subdued by police. Thirteen deaths that became part of Westlake's dark history.

She glanced at Heather, at her terrified expression, and felt that strange surge of protectiveness again. What if Heather had been one of the dead that night—this night? What if Sienna could change that?

"Look," Sienna said, turning to face the group. "I know this sounds crazy, but I think if we work together, we might be able to stop him."

The pirate laughed, a harsh sound with no humor. "Stop him? With what? Our costume props?"

Sienna assessed their surroundings. The bathroom offered little in terms of weapons. The metal trash cans could be used for battering but were unwieldy. They could break the mirror and gather fragments for stabbing.

"There's five of us and one of him," she argued. "If we coordinate, create a distraction—"

"Four," Princess Leia corrected, pointing at Danny's lifeless form. "There's four of us left."

Heather studied Sienna with newfound intensity. "How do you know so much about what to do?"

The question caught Sienna off-guard. How could she explain that she knew the story of tonight from history and local legends? That in her time, this night was infamous?

"I just . . . I know we can't wait here to die," she said, avoiding the impossible explanation.

A scream, closer than the others, echoed down the hallway. The sound of running footsteps passed the bathroom door, followed by a heavy thud and the wet sound of a blade finding flesh. The footsteps transitioned from frantic to halting, then a body slumped against the wall outside, sliding down to the floor with a muffled thump.

Princess Leia clapped her hands over her ears, tears streaming down her face. "We're all going to die," she whispered, rocking slightly.

Sienna approached the mirror, studying her reflection. Madonna costume splattered with blood, blonde hair disheveled, fear etched across her features. With sudden determination, she drove her elbow into the mirror, shattering it into large shards that crashed into the sink. She picked up the largest piece, wrapping one end with paper towels to create a makeshift handle.

"I'm going out there," she announced, the glass shard glinting under the fluorescent lights. "I'm going to find a better weapon, and I'm going to stop this."

"You'll get yourself killed," the pirate warned.

"Maybe," Sienna conceded. "But I can't just hide and wait for it to happen." She looked at Heather, feeling a strange responsibility toward her—this girl who, in another timeline, might not have survived the night. "You should stay here. Barricade the door after I leave."

Heather's expression shifted between fear and determination, her blonde hair glowing almost white under the harsh bathroom lights. Blood—Danny's blood—had spattered across her Madonna costume, a grotesque mirror to Sienna's own.

"Be careful," she said finally.

Sienna nodded, helping the others move the barricade enough for her to slip through. The hallway beyond was dimly lit, emergency lights casting long shadows that seemed to move of their own accord.

"Lock it behind me," she instructed, gripping her mirror shard tightly. She stepped into the hallway, hearing the immediate scrape of the barricade being replaced.

ALONE IN THE CORRIDOR, Sienna took a deep breath. The school had transformed into something nightmarish. Blood smeared the lockers in long, dragging streaks where wounded students had leaned for support. A trail of bloody footprints led away from a dark pool that had formed outside the bathroom—evidence of the killing they'd just heard.

At the end of the hall lay a crumpled form, facedown and motionless. The victim's costume—some kind of superhero—was now so saturated with blood that its original colors were indiscernible. One arm was splayed outward, fingers curled as if still reaching for escape. The back of the victim's head was partially caved in, dark matter and bone fragments scattered on the linoleum.

"Thirteen victims," Sienna whispered to herself, remembering the story. "Let's change that number." She moved cautiously down the hallway, mirror shard held like a knife, every sense straining for signs of the killer.

The school had never felt so vast or so threatening. Each shadow might hide Teddy, each closed door might conceal another tragedy. The only sounds were her own careful footsteps and the distant, muffled cries of other students hiding or fleeing. She moved down the corridor with careful steps, her white heels making minimal noise on the linoleum floor. The emergency lighting cast everything in an eerie red glow, transforming the familiar hallways of Westlake High into something from a nightmare.

She passed a bulletin board with Halloween decorations and student announcements. A paper advertised "HALLOWEEN DANCE—OCTOBER 31, 1984" with cartoon ghosts and pumpkins that now seemed like a sick joke.

Rounding a corner, Sienna encountered another body—a girl in a cat costume sprawled facedown, a dark pool spreading beneath her. The girl's hand was still clutching a plastic pumpkin candy bucket, now spilled across the floor, its contents mingling with the blood. The back of her neck gaped open, vertebrae visible through the severed flesh. The sight made Sienna's stomach lurch, but she forced herself to keep moving.

This wasn't just a story anymore—it was happening in real-time, and the victims were real teenagers with their whole lives ahead of them. Or should have been.

Blood trails led in different directions, telling stories of desperate escapes and failed attempts at hiding. Handprints marked the walls where wounded students had tried to steady themselves. The metallic smell of fresh blood filled her nostrils, making it hard to breathe.

A classroom door stood ajar. Sienna peered inside to see chairs overturned and papers scattered. A teacher's body lay slumped over the desk, the back of her head a ruined mess of hair, bone, and tissue. The machete had cleaved through the skull with horrible efficiency.

The mirror shard in Sienna's hand suddenly seemed woefully inadequate. She scanned the hallway for something better.

An emergency case on the wall caught her eye—red with a glass front, containing a fire axe. The words BREAK GLASS IN CASE OF EMERGENCY had never seemed more appropriate.

Sienna used her elbow to shatter the glass, the sound unnervingly loud in the quiet hallway. She froze, listening for approaching footsteps, for the telltale sound of Teddy's machete dragging against the lockers.

Hearing nothing, she carefully removed the axe, testing its weight in her hands. The solid heft of it was reassuring compared to the fragile mirror shard, which she discarded on the floor. The axe handle was smooth, well-maintained—designed to be functional rather than merely decorative.

A distant scream echoed through the school, followed by the sound of running footsteps. Sienna pressed herself against the wall, axe ready, her heart hammering so loudly she was sure it could be heard down the corridor.

Two students—a boy and girl in matching cowboy costumes—sprinted around the corner, terror evident on their faces. They stopped abruptly when they saw Sienna with the axe, the girl stumbling into the boy.

"This way!" the boy gasped, pointing back the way they came. "He's right behind us!"

Before Sienna could respond, a steady thudding sound approached—the sound of someone walking unhurriedly, confident in their pursuit. The measured pace was somehow more terrifying than a frantic chase would have been.

"Run," Sienna told them, raising the axe. "Find somewhere to hide."

The couple hesitated for only a moment before continuing their flight down the hallway. Sienna turned to face the approaching threat, gripping the axe tightly, its weight suddenly seeming insufficient.

Around the corner appeared Teddy Kopeki, his clown mask now spattered with blood. The white face paint and exaggerated red smile created a grotesque contrast with the very real machete in his hand, its blade darkened with drying blood.

Teddy stopped when he saw Sienna, his head tilting to one side like a curious animal. The eye holes of the mask revealed nothing of the person beneath, but Sienna felt his gaze assessing her, measuring the threat she might pose with the axe.

"Teddy," she said, her voice surprisingly steady. "You don't have to do this."

The masked figure remained motionless for a long moment. There was something unnerving about his stillness—no fidgeting, no nervous energy, just cold focus. The only movement was the slow drip of blood from the machete onto the floor, creating a soft, rhythmic patter.

Then, with unexpected speed, he turned and disappeared down a side corridor—not retreating, but changing course to pursue easier prey.

Sienna realized he was going after the cowboy couple. She followed, trying to move quietly despite her heels. She kicked them off, the cool linoleum against her bare feet allowing for silent movement as she tracked Teddy through the darkened school.

At each intersection, she paused, listening. The school had fallen eerily quiet, the kind of silence that felt heavy with threat. Distant whimpers and occasional movement were the only signs that others still survived.

A muffled sob drew her attention to the AV room ahead. The door was slightly ajar, a thin strip of light visible from inside. Approaching carefully, Sienna peered through the crack. Several students were huddled inside, including the cowboy couple. They'd barricaded the door with an AV cart and were trying to stay quiet.

Sienna was about to knock when movement at the end of the hall caught her eye. Teddy stood silhouetted against the emergency lights, machete hanging at his side, watching the AV room with predator's patience.

She ducked into an empty classroom across the hall, leaving the door cracked. Through this vantage point, she watched as Teddy approached the AV room with unhurried steps, like a hunter who knows his prey is cornered.

Without warning, Teddy kicked the door open, sending the AV cart crashing backward. The barricade was never going to hold against such force—it had only provided the illusion of safety.

Screams erupted from inside. From her hidden position, Sienna saw the scene unfold in horrific flashes as a film projector inside the room was knocked over, its beam swinging wildly across the space.

In strobe-like bursts of light: Teddy advancing into the room. The cowboy raising his hands defensively. The machete swinging in a gleaming arc. The cowboy's neck opened in a fountain of red, his hat flying off as he collapsed.

His girlfriend's scream cut through the air, primordial and raw. More flashes: Teddy moving methodically to his next victim. A boy in a Ghostbusters costume tried to fight back with a chair. The machete came down, splitting his shoulder to sternum in a diagonal slash. Blood sprayed across the film projector lens, casting the room in a red filter.

Sienna's muscles tensed as adrenaline flooded her system. She'd spent years in kickboxing classes—a hobby her friends had teased her about. Now it might save her life. She gripped the axe firmly, measuring the distance to Teddy, calculating the angle of attack.

In the swinging projector light, she saw the cowgirl

crawling desperately toward the door, leaving a smear of blood behind her. Beyond her, a figure in white—another Madonna costume similar to Sienna —was huddled in the corner.

Teddy stepped over the cowgirl, focused on the Madonna costume. Sienna's heart stopped, thinking it was Heather, before realizing it was another student.

Sienna burst from her hiding place, axe raised. "Hey!" she shouted, wanting to draw Teddy away from the trapped students.

Teddy turned, the clown mask looking absurdly cheerful splattered with fresh blood. For a moment, he seemed genuinely surprised by her intervention.

Sienna swung the axe, aiming for his shoulder. Teddy sidestepped with unexpected agility, the axe blade embedding in the wooden doorframe with a solid thunk. The impact sent shock waves up her arms, nearly causing her to lose her grip.

Refusing to let go, Sienna braced her foot against the wall and yanked the axe free. Teddy lunged with his machete, forcing her to dodge backwards. The blade sliced through the air where her abdomen had been moments before.

Behind the mask, eyes assessed her with cold intelligence. This wasn't mindless rampage—there was calculation in his movements, a deliberate quality that made him all the more terrifying.

Sienna circled right, keeping her weight balanced as she'd learned in kickboxing. Teddy matched her movements, machete held at the ready. He was waiting for her to make a mistake.

The Madonna-costumed girl used this distraction to scramble toward the door. Teddy noticed, pivoting to catch her. Sienna seized the opening, driving forward with the axe. This time, she caught him across the back of his arm, the blade biting into flesh with a meaty thunk.

Teddy made no sound despite the wound. He turned back to Sienna, examining the gash on his arm with what seemed like curiosity rather than pain. Dark blood soaked through his jacket sleeve.

Sienna advanced, pressing her advantage. She feinted left, then swung the axe in a tight arc as Teddy moved to block. The blade connected with his side, cutting through his jacket and into his ribs with a crunch of bone.

He staggered but didn't fall. With shocking speed, he slashed the machete upward. Sienna jerked back, but not fast enough. The blade sliced across her forearm, opening a four-inch gash that immediately welled with blood.

Pain lanced up her arm, hot and immediate. Sienna switched the axe to her left hand, backing up to create distance. Blood ran down her injured arm, dripping from her fingertips onto the floor.

In the flickering projector light, Teddy's mask shifted, revealing a sliver of teenage face beneath. One human eye locked onto hers—cold, calculating, but unmistakably human. The sight was somehow more terrifying than the expressionless mask. This wasn't a monster from a horror movie; this was a teenager making conscious choices to slaughter his classmates.

Sienna kicked over a desk, creating a momentary barrier. Teddy stepped around it with predatory grace, but the interruption gave her time to adjust her grip on the axe. Her wounded arm throbbed, each heartbeat pushing more blood through the gash.

The Madonna-costumed girl had reached the door. "Run!" Sienna shouted to her, never taking her eyes off Teddy.

Teddy lunged again, machete sweeping in a horizontal arc. Sienna ducked under the blade, feeling it whistle over her head, then drove upward with her axe. The blade caught Teddy's wrist, severing tendons and chipping bone.

Blood sprayed across the room as the machete clattered to the floor, his hand suddenly useless. For the first time, Teddy made a sound—a surprised grunt that revealed his vulnerability.

Sienna again pressed her advantage, driving her shoulder into his chest and sending him crashing into the film equipment. Teddy stumbled but recovered quickly, reaching for his fallen machete with his uninjured hand.

She brought the axe down hard, aiming for his reaching arm, but he rolled aside. The blade struck the floor, sending painful vibrations up her arms. Before she could lift it again, Teddy's boot connected with her stomach, driving the air from her lungs.

Sienna doubled over, momentarily defenseless. Teddy retrieved his machete, switching it to his left hand. Blood dripped steadily from his mangled right wrist, leaving a trail across the floor.

She forced herself upright, fighting for breath. They circled each other again, both wounded, both reassessing. The Madonna girl had escaped, along with the cowgirl. Two lives saved.

The projector beam swung wildly, creating disorienting shadows that made tracking Teddy's movements difficult. He attacked with unexpected speed, the machete slashing toward Sienna's face. She jerked backwards, the blade missing her eyes by inches but catching her cheek, leaving a shallow cut that immediately stung with sweat.

Blood trickled down her face as she countered, swinging the axe in a controlled arc. Teddy backpedaled, but his injured leg slowed him. The axe blade caught his shoulder, sinking into flesh, stopped only by his collarbone.

He wrenched away, tearing the axe from her grip. Both weapons clattered to the floor, leaving them momentarily unarmed. Teddy lunged forward, his greater weight and height an advantage in close combat. His hands closed around her throat, thumbs digging into her windpipe.

Sienna drove her knee up, aiming for his groin but connecting with his thigh instead. His grip loosened just enough for her to twist sideways, breaking his hold. She gasped for air, spots dancing at the edges of her vision.

Teddy reached for the fallen machete. In desperation, Sienna grabbed a film projector and heaved it at him. The heavy equipment struck his chest, sending him staggering into the wall. Glass shattered, metal components scattering across the floor.

She scrambled for the axe, fingers closing around the handle just as Teddy recovered. He had his machete now, blood dripping down his arm, his breathing audible through the damaged mask—labored but controlled.

"Who are you?" he asked, his voice startlingly normal, a teenager's voice. It was the first time he'd spoken.

The question caught Sienna off-guard. Before she

could answer, footsteps thundered in the hallway—more students fleeing, their panic audible even from a distance.

Teddy's head turned toward the sound, the predator in him drawn to easier prey. He backed toward the door, keeping the machete between them.

"This isn't over," he said, then slipped into the hallway.

SIENNA COLLAPSED AGAINST THE WALL, her body suddenly acknowledging every injury. Her forearm throbbed, blood seeping through the torn lace of her costume. Her throat felt bruised, each breath painful. The cut on her cheek burned.

But she was alive. And she had to keep moving.

She forced herself upright, retrieving the axe. Outside in the hallway, she heard Teddy's footsteps receding—not running, never running, just that measured, confident walk.

The lights in the AV room flickered, casting shadows across the bodies of the cowboy and the Ghostbusters fan. Their vacant eyes stared at nothing, blood pooling beneath them on the tiled floor.

Sienna followed the sound of Teddy's footsteps, favoring her uninjured side. The halls were eerily quiet now—no more screams, no more running. Either everyone had escaped, hidden, or...

She refused to complete the thought.

Rounding a corner, she caught sight of Teddy at the far end of the corridor. He stood outside the bathroom where she'd left Heather and the others, studying the door with that same predatory patience. His hand reached for the handle.

"Hey!" Sienna shouted, forcing strength into her voice despite her bruised throat. "I'm not finished with you yet!"

Teddy turned, the machete gleaming under the emergency lights. Blood covered his left side where she'd struck him, but he showed no signs of weakening. If anything, the injuries seemed to have focused him, stripped away any remaining hesitation.

Sienna advanced, axe ready. Her mind raced, calculating the distance between them. She needed to draw him away from the bathroom, away from Heather and the others.

"You asked who I am," she called to him, continuing her approach. "I'm the one who's going to stop you, Teddy."

His head tilted at the sound of his name, that same

animal curiosity. The mask had shifted further, revealing half his face now—a teenager with sharp features, blood spattered across his cheek, one eye cold and evaluating.

"You don't belong here," he said, voice flat.

Sienna faltered for a fraction of a second. How could he tell? Was there something about her that didn't fit this time, this place?

She recovered quickly, continuing her approach. "Put down the machete, Teddy. This doesn't have to end the way you think."

"It's already ended," he replied with chilling certainty. "For them." He gestured with the bloody machete toward the trail of carnage behind him.

Teddy abandoned the bathroom door, moving toward Sienna with that same measured pace. His left hand wielded the machete with surprising skill, compensating for his mangled right.

"Why are you doing this?" Sienna asked, partly to stall, partly out of genuine curiosity about the motivation behind the infamous massacre she'd only heard whispered about decades later.

"They deserve it," he said simply. "All of them. Watching, laughing, pretending not to see." His face— what was visible of it—remained eerily expressionless, as though discussing the weather rather than murder.

Sienna backed down the corridor, drawing him away

from the bathroom. She needed a better battleground, somewhere with more room to maneuver.

"Not everyone deserves to die," she countered, keeping her voice level. "Not Heather. Not the ones hiding."

At Heather's name, something flickered in Teddy's visible eye—recognition, perhaps even emotion. "Especially *her*," he said, voice hardening. "Queen of them all, isn't she? Always perfect, always *untouchable*."

They reached an intersection in the hallway. Sienna backed toward the cafeteria, its double doors visible at the end of the corridor. More space there, more potential weapons.

"I won't let you hurt anyone else," she said, tightening her grip on the axe.

Teddy lunged suddenly, the machete slashing toward her midsection. Sienna pivoted, the blade cutting through her dress but missing flesh. She countered with a swing of the axe that Teddy barely avoided, the blade catching his jacket sleeve and tearing the fabric.

They crashed through the cafeteria doors, Sienna backing away as Teddy pursued. The large room with its long tables offered more maneuverability, but also fewer places to hide. Moonlight filtered through high windows, casting everything in silver and shadow.

Sienna circled around a table, keeping it between them. Blood from her arm wound had soaked through to her elbow, making the axe handle slippery in her grip. She switched hands, wiping her palm against her dress.

"You can't save them," Teddy said, stalking her with predatory focus. "You don't even belong in this time."

The statement hit her like a physical blow. He knew. Somehow, he sensed the truth about her.

"What do you mean?" she asked, buying time as she assessed her surroundings.

"I see things others don't," he replied, tapping his temple with his injured hand. Blood had caked the entire right side of his jacket now, yet he moved with undiminished precision. "You're from somewhere else. Some*when* else."

He charged suddenly, machete raised. Sienna dodged behind a milk dispenser, which exploded in a spray of white liquid as Teddy's blade cut through it. The machine crashed to the floor, dairy splashing across the tiles.

Sienna used the distraction to close the distance, swinging the axe at Teddy's exposed side. He pivoted, but not quickly enough—the blade bit into his hip, eliciting the first genuine sound of pain from him, a strangled gasp.

He staggered backwards, free hand clutching the new

wound. Blood seeped between his fingers, darker than the milk spreading across the floor.

Sienna advanced with the axe raised. Teddy stumbled against a serving counter, momentarily cornered.

"It's over, Teddy," she said, keeping the axe leveled at his chest. "Drop the machete."

Instead, he lunged desperately, slashing wildly. The blade caught Sienna's shoulder, slicing through dress and skin. She cried out, the sudden pain nearly causing her to drop the axe.

Blood immediately soaked through the white lace of her costume, hot and pulsing. The cut wasn't deep, but it burned like fire, her arm threatening to go numb from shock.

Teddy stumbled toward the kitchen doors, leaving a trail of blood behind him. Sienna followed, determined not to let him escape. She pushed through the swinging doors into the industrial kitchen with its stainless steel counters and large appliances. The room smelled of that day's lunch—spaghetti and meatballs, according to the menu board—mixed with the more immediate scent of fear and blood.

"There's nowhere left to run," she called out, scanning the dimly lit kitchen. Her voice echoed off the tiled walls.

A shape lunged from behind an industrial refrigerator. Teddy swung the machete downward with desperate strength. Sienna raised the axe to block, the blades meeting with a metallic screech. The impact drove her back a step, her wounded shoulder screaming in protest.

Teddy forced her backward until she collided with a rack of pots and pans. They crashed to the floor, the cacophony deafening in the tiled kitchen.

Sienna seized a heavy cast iron skillet, hurling it at Teddy's head. He deflected it with the machete but lost his balance slightly.

She followed with a swift kick to his injured hip. Teddy buckled, dropping to one knee with a grunt of pain. The clown mask slipped further, revealing most of his face now—pale, sweat-streaked, teeth gritted in determination.

"Why fight so hard?" he gasped, blood bubbling at the corner of his mouth, suggesting internal injuries. "You don't know these people. They were dead long before you came."

"Not all of them," Sienna replied, thinking of Heather. "And not tonight."

Teddy lunged again, his movements growing erratic. The machete slashed wildly, catching Sienna's thigh as she tried to dodge. She felt the blade slice through skin, a painful but relatively shallow cut. Her leg faltered momentarily from the shock.

Gritting her teeth against the searing pain, Sienna braced herself against a counter to regain her balance. Blood trickled down her leg, staining her torn dress. Every instinct screamed at her to flee, to find safety and stem the bleeding, but she knew what would happen if she let Teddy escape.

He circled her, machete dripping with her blood. Despite his own injuries, he sensed her weakness, moving to position himself between her and the exit.

"History will remember me," he said, his voice eerily calm despite the blood trickling from his mouth. "They'll talk about tonight for decades. You know that, don't you? That's why you're here."

Sienna spotted a container of cooking oil on a nearby shelf. She lunged for it, ignoring the white-hot pain in her leg, and threw it at Teddy's feet. The oil spilled across the floor, creating a hazardous slick.

"History remembers the victims," she countered, adjusting her grip on the axe. "Not you."

Rage flashed across Teddy's visible features. He charged, slipping slightly in the oil but maintaining enough balance to swing the machete at her head. Sienna ducked, feeling the blade whistle over her hair.

She pivoted on her good leg and swung the axe with

all her remaining strength. The blade caught Teddy's forearm, severing tendons and arteries in a spray of blood that splattered across the stainless steel counters like macabre abstract art.

The machete clattered to the floor, his hand suddenly unable to grip it. Teddy stared at his useless arm in shock, arterial blood pulsing from the wound in rhythmic spurts.

He staggered backward, slipping in the mixture of oil and blood. For a moment, his eyes met Sienna's—confusion and pain replacing the cold certainty that had been there before.

"You can't change what happens," he said, his voice weaker now. "This night has already occurred."

"We'll see about that," Sienna replied, retrieving the fallen machete with her free hand. The weight of it felt obscene, knowing what it had done tonight, whose blood stained its edge.

Teddy's eyes widened at the sight of her holding both weapons. He fumbled behind him, reaching for something on the counter. His hand closed around a large kitchen knife, raising it with desperate determination.

"You don't understand," he said, blood bubbling between his lips. "I've seen this night in dreams for years. Always the same. Always thirteen dead."

He lunged forward, knife aimed at her throat. Sienna twisted away, her injured leg nearly collapsing beneath her. The knife caught her upper arm instead, adding another line of fire to her collection of wounds.

She brought the machete down in a sweeping arc, feeling it connect with bone as it bit deep into Teddy's shoulder. He screamed as the blade lodged in his collarbone.

Teddy staggered back, the machete still embedded in his shoulder. Blood soaked his entire left side now, his breathing ragged and wet. He careened into the large steel refrigerator, leaving a smear of crimson across its surface.

"It can't end like this," he gasped, sliding down to a sitting position, leaving a trail of blood on the refrigerator door. "Not with me. I'm supposed to..."

His voice trailed off as shock set in, his eyes growing unfocused. The kitchen knife fell from his limp fingers, clattering on the tile.

Sienna approached cautiously, axe ready. Teddy's gaze drifted up to meet hers, clarity returning briefly through the haze of blood loss.

"Who are you?" he whispered, the question genuine now, stripped of menace.

"Someone who's changing history," she answered, raising the axe for what might be a final blow.

Teddy's eyes widened, a flicker of confusion crossing his face. "This isn't . . . how it's supposed to end," he mumbled, blood bubbling at the corner of his mouth.

Before he could say anything else, a noise from the doorway caught Sienna's attention. Heather stood there, clutching a fire extinguisher like a weapon, her Madonna costume bloodied and torn, her eyes wide with determination and fear.

"Sienna!" she called out.

Teddy's head whipped around at the sound, his body suddenly tensing with renewed purpose. Despite his grievous injuries, he lunged for the fallen kitchen knife, his movements fueled by a final surge of adrenaline.

"No!" Sienna shouted, charging forward. The axe connected with Teddy's back, sinking between his shoulder blades with a sickening thud.

Teddy arched backward, a choked gasp escaping his lips. The knife fell from his suddenly nerveless fingers. He made a wet, rattling sound, reaching behind himself futilely with one hand.

Heather stood frozen, watching as Teddy stumbled forward, then crashed to his knees. Blood bubbled from his lips as he struggled to breathe, his eyes wide with shock more than pain.

Sienna retrieved the fallen machete, standing over Teddy with both weapons now. "It's over," she said, her voice shaking with adrenaline and exhaustion.

Teddy looked up at her, recognition flickering in his

dying eyes. "You..." he whispered, blood painting his lips crimson. "You're not . . . supposed to be here."

His body pitched forward onto the blood-slicked floor. His fingers twitched once, twice, then stilled completely. Teddy Kopeki, the Halloween Massacre killer, lay dead in a growing pool of his own blood.

Sienna and Heather stared at each other across his body, both spattered with blood, both struggling to process what had happened. For a moment, there was only the sound of their ragged breathing and the steady hum of the refrigeration units.

Then the fluorescent kitchen lights flickered violently, plunging the room into darkness, then light, then darkness again. A strange humming filled the air, growing louder with each flicker.

"What's happening?" Heather gasped, reaching for Sienna's hand across the darkness.

"I don't know," Sienna answered truthfully, feeling a strange pulling sensation, like gravity shifting direction. The Madonna dress seemed to tighten around her, growing hot against her skin.

The kitchen tilted and spun, reality itself seeming to warp. The last thing Sienna saw was Heather's terrified face illuminated in strobe-like flashes as the world dissolved around them.

MUSIC HIT SIENNA FIRST—NOT the 80s tracks from the Halloween dance, but modern remixes of classic songs. The kind played at reunion parties.

Disoriented, she blinked against suddenly bright lights. The gymnasium swam into focus around her, decorated with "Welcome Back Class of 2004" banners and photo collages of smiling teenagers who were now approaching middle age.

People in costumes danced around her, many wearing 80s-themed outfits, but these were adults in their late thirties, not teenagers. Her classmates, twenty years older than when she last saw most of them.

The axe and machete were gone from her hands. She looked down at herself, finding her Madonna costume torn and bloody, exactly as it had been moments ago in 1984.

"What the hell?" someone nearby exclaimed. The music continued but the dancers closest to Sienna had stopped, staring at her blood-soaked appearance.

Sienna touched her arm—the cut from Teddy's machete was still there, still bleeding. This wasn't a dream or hallucination; the physical evidence remained.

Jordan pushed through the gathering crowd. "Si? Jesus Christ, what happened to you?" His face was twisted with concern.

"I..." Sienna started, but had no explanation that wouldn't sound insane. She looked around wildly, trying to orient herself in time and space.

"Someone call an ambulance," a woman shouted, pointing at Sienna's bleeding arm.

"No," Sienna said, finding her voice. "I'm okay. It's . . . it's fake. Part of the costume." The lie sounded hollow even to her own ears, but she needed time to process what had happened.

Jordan didn't buy it. "That's real blood, Si. What the hell is going on? Did someone attack you?"

Before she could answer, the lights dimmed further. The DJ spoke into his microphone: "And now, as promised, let's take a trip down memory lane with our Westlake High retrospective slideshow!"

A projector sprang to life, casting images onto the white wall behind the DJ booth. Photos of their graduation, sporting events, and candid moments from their high school years appeared, accompanied by nostalgic music.

The crowd's attention shifted to the slideshow, momentarily forgetting Sienna's bizarre appearance. Jordan stayed beside her, his hand on her uninjured arm.

"Let me take you to the bathroom," he said quietly. "We need to clean those cuts."

Sienna shook her head, her eyes fixed on the slideshow. "Wait. Just wait." Something told her to watch, to see what came next.

The slideshow transitioned to "Westlake Through the Years," showing historical photos of the school and important events. One slide read "Remembering the Halloween Tragedy—1984."

Sienna's breath caught. The next slide showed a newspaper headline: "Seven Dead in Halloween Attack—Student Hero Stops Massacre"

Photos appeared—yearbook photos of the victims, and then the group photo taken earlier that night. Heather, her friends, and unmistakably, Sienna in her Madonna costume, standing among the students from 1984.

Jordan's grip on her arm tightened painfully. "What the fuck, Si?" he whispered, staring at the impossible image. "Is that actually you in that picture?"

Others had noticed too. Heads turned toward Sienna, confusion and disbelief on their faces. The photo was clearly decades old, yet she stood among students from 1984, wearing the exact same costume she wore now.

The slideshow continued, but Sienna barely noticed. Reality seemed to waver around her as she processed what had happened—she had changed history, altering a tragedy that had always been part of Westlake's past.

"I need some air," she said to Jordan, pulling away from his grip. She pushed through the crowd toward the exit, leaving bloody handprints on the doors as she burst outside into the cool October night.

The pain from her wounds hit her fully now, the adrenaline no longer masking it. She leaned against the brick wall, breathing hard, trying to make sense of everything.

Jordan burst through the doors moments later, a dampened cloth in his hand. "Jesus, Si, you're really hurt," he said, pressing the cloth against her bleeding arm. Behind him, others from the reunion gathered, someone calling 911, another bringing a first aid kit from the office.

"I'm fine," she insisted weakly, but the world was beginning to spin.

"The hell you are," Jordan said, his voice firm but gentle. "You need a hospital. This isn't fake blood, and these aren't costume wounds."

Sienna wanted to argue, but found she didn't have the strength. As the distant wail of an ambulance approached, she finally surrendered to the pain and confusion, slumping against Jordan's supportive shoulder.

A WEEK AFTER THE REUNION, Sienna sat at her kitchen table, laptop open to news archives. Her arm—the worst of the many wounds Teddy had gifted her—was properly bandaged now. The gash—that she explained away as an 'accidental run-in with a mirror'— required twenty stitches. The doctor had looked skeptical but hadn't pressed further. Her thigh wound had been less severe than it felt in the moment—though it still throbbed when she moved too quickly. The cut on her face was superficial and would likely heal with minimal scarring.

The screen displayed a newspaper article from November 1st, 1984: "Halloween Horror at Local High School." She read for the dozenth time about how Teddy Kopeki killed seven students before being stopped by "an unidentified female student" who disappeared in the chaos.

In Sienna's original timeline, thirteen students had died. Now history recorded only seven victims—she had changed the past, saved six lives. Mike wasn't among them. His death still appeared in the article, along with a senior named Patricia Griffith and a junior named Aaron

Wellman. But the cowboy and third Madonna had survived, as had the girl in the prom queen costume.

She scrolled through more articles, finding a follow-up piece about the survivors. Heather Morgan was mentioned specifically—quoted about her narrow escape and her gratitude to the mysterious student who saved her life.

"I don't know who she was," seventeen-year-old Heather had told the reporter. "She said she was visiting relatives. But she knew exactly what to do, and she saved our lives."

Sienna typed "Heather Morgan Westlake" into the search bar. A professional website appeared at the top of the results. Heather, now in her fifties, had become a trauma counselor specializing in helping survivors of violence. Her practice was located just twenty minutes from Sienna's apartment.

The website showed a photo of Heather—silver-blonde hair, kind eyes with smile lines, professional yet approachable. Forty years had passed, but Sienna could still see the teenager she'd met on Halloween night 1984.

She closed the laptop, mind racing with implications. Had she truly altered history? Or had this always happened, a closed time loop where she was always meant to go back?

Sienna walked to her closet, pulling out the garment

bag containing the ruined Madonna costume. She hadn't been able to bring herself to throw it away, despite the blood stains that had dried to a rusty brown. The white lace was torn in numerous places where Teddy's machete had shredded it, and the "Boy Toy" belt had a crack running through the B where it had deflected a blow meant for her abdomen.

As she unzipped the bag to examine the dress one last time before disposal, something fluttered to the floor—a business card.

Heather Morgan, LMFT Trauma Recovery Specialist

The address matched the one from the website.

Sienna turned the card over. On the back, handwritten in blue ink: "Some costumes change more than just appearances. Thank you for that night." It was signed "Heather Morgan, 1984."

Her hands trembled as she stared at the card. How had Heather known? Did she recognize Sienna all these years later? Or did she somehow understand what happened that night?

She tucked the card into her pocket and returned to the dress. As she examined it one last time, she noticed a small tag she hadn't seen before, sewn into the lining. The text was faded but legible: "Property of Westlake High Drama Dept., 1984."

The business card hadn't been there before—she was

certain of it. But then, neither had the tag identifying the dress as 'Property of Westlake High Drama Dept., 1984.' It seemed that altering history had altered the dress too, leaving behind evidence of the new timeline she'd created.

"Second Chances," Sienna whispered, remembering the name of the shop. The elderly shopkeeper's words echoed in her mind: "That piece has been here for years. No one's shown interest in it before."

The dress had created a loop through time—Heather survived to donate the dress that would eventually save her life. The paradox made Sienna's head hurt if she thought about it too hard.

She carefully returned the dress to the garment bag and slipped it back into the closet. Then she picked up her phone. On the screen was Heather Morgan's office number from the business card.

Sienna's finger hovered over the call button as she considered what to say to a woman she both knew intimately and had never properly met. How do you start a conversation about a shared trauma that happened both forty years ago and just last week?

After a moment's hesitation, she pressed call. As the phone rang, Sienna touched the scar forming on her arm —physical proof of impossible events, a permanent reminder of a night that changed history.

"Morgan Trauma Recovery," a receptionist answered. "How may I help you?"

Sienna took a deep breath. "I'd like to speak with Heather Morgan." Her voice was steadier than she expected. "Tell her . . . tell her it's about a Halloween costume we once shared."

Sienna waited on hold, flicking the business card between her fingers. The picture of Heather on the website stared back at her—older, wiser, but still recognizable as the brave teenager who'd welcomed a stranger on Halloween night.

ABOUT THE AUTHOR

Steven Pajak, a Chicago-based author, crafts stories that explore the depths of horror and the human psyche. With a pen that dances on the edges of darkness, Steven brings to life tales that challenge, terrify, and linger in the minds of readers. Drawing inspiration from the urban tapestry of Chicago, his work merges the pulse of city life with the eerie quiet of the shadows lurking within the darkest corners of our minds. Steven invites you into a world where fear meets courage, and the journey through his imagination proves as haunting as it is unforgettable.

facebook.com/StevenPajakAuthor

instagram.com/stevenpajak_official

amazon.com/author/stevenpajak

goodreads.com/stevenpajakauthor

www.ingramcontent.com/pod-product-compliance
Lightning Source LLC
Chambersburg PA
CBHW010959010826
48969CB00016BA/2633